WHAT IS
PHILOSOPHY FOR?

ALSO AVAILABLE FROM BLOOMSBURY

A History of Animals in Philosophy, Oxana Timofeeva
Environmental Ethics, Marion Hourdequin
Can't We Make Moral Judgements? Mary Midgley
Why Iris Murdoch Matters, Gary Browning

WHAT IS PHILOSOPHY FOR?

BY MARY MIDGLEY

BLOOMSBURY ACADEMIC
LONDON • NEW YORK • OXFORD • NEW DELHI • SYDNEY

BLOOMSBURY ACADEMIC
Bloomsbury Publishing Plc
50 Bedford Square, London, WC1B 3DP, UK
1385 Broadway, New York, NY 10018, USA

BLOOMSBURY, BLOOMSBURY ACADEMIC and
the Diana logo are trademarks of Bloomsbury Publishing Plc

First published in Great Britain 2018
Reprinted 2018 (twice), 2019

Cover design: Irene Martinez-Costa
Cover image: Butterflies, moths and other insects with a sprig of apple blossom
and periwinkle (oil on copper), Kessel, Jan van, the Elder (1626–79) © Johnny Van
Haeften Ltd., London / Bridgeman Images

A catalogue record for this book is available from the British Library.

A catalog record for this book is available from the Library of Congress.

ISBN: HB: 978-1-3500-5108-9
        PB: 978-1-3500-5107-2
        ePDF: 978-1-3500-5109-6
        eBook: 978-1-3500-5110-2

Typeset by Integra Software Services Pvt. Ltd.
Printed and bound in Great Britain

To find out more about our authors and books visit www.bloomsbury.com
and sign up for our newsletters.

CONTENTS

PART ONE

THE SEARCH FOR SIGNPOSTS

1

Directions

What is the aim, the proper object of philosophizing? What are we trying to do?

We are not, of course, starting from nowhere, nor are we just riffling through ideas at random. We are always looking for something particular – a link, a connection, a context that will make sense of our present muddled notions. Thoughts that are now blundering around loose and detached need somehow to be drawn into a pattern – perhaps eventually into a single pattern, an all-explaining design. And when one of these patterns seems to be becoming more complete, we approve, and begin to speak of it hopefully as a *philosophy*.

But this doesn't always work.

Often we seem to be trying to resolve a complex jigsaw, one which has mistakenly brought together parts of several different pictures; trying to give a single shape to a manifold vision. Indeed, we are bound to keep doing this, because our minds are never quite empty at the start. They always contain incomplete world-pictures, frameworks to which loose scraps of experience and of various studies, such as geology, history, mathematics, astronomy and so forth, can be attached. And these various frameworks do not fit together spontaneously. They have usually grown from different sources, sometimes from distinct social groups which seem to us to be alien, even hostile to one another.

In this way people educated as Muslim or Christian literalists may not be able to find any place in their minds for the theories of modern physics, although those theories were devised in the first place by people who were devout believers. And up-to-date physicists may be just as mystified about where to put other people's religious ideas ... Moreover, even within our own branch of religion or of science, each of us can be confronted once more by division – by the need to choose. Quite possibly, indeed, what made us start our questioning in the first place was a contemporary debate, a puzzle which offers only two alternative solutions. And, since we know that scientists are meant to make Discoveries, we want to arrive now at a Discovery, a single final answer.

The quest for quanta

Thus (again) the standing difficulty about quantum theory arises from the clash between two different ways of interpreting it. Should we think in terms of waves or particles? We ask, which of these images is the better? The reply to this is not simple. It depends on a mass of wider calculations, among which we need to choose – not just a single preferred language, but a whole wider background, a world-picture.

This trouble arises even if we only ask modest questions, such as: is the world flat or round? Though today we are supposed to know that it is round, we never see it that way. And if we accept that it actually is round, we have to ask endless further questions about why it is so, which involve the nature and position of the rest of the universe, indeed of Nature itself. Since we cannot isolate any one of these questions, we are not just choosing between the two alternatives envisaged by today's disputants. Instead, we are trying to discover the nature of the earth as a whole – this multiformed planet, including all its insects, its birds, its fish, its humans and the secrets of its core. And we are not just asking about these things from outside, as mere neutral observers, but as creatures concerned about them because we are included within them. In particular today, we have to find the right place for scientific suggestions about all this on the larger map of life as a whole.

This is why philosophy does not 'progress' in a straight line, adding one discovery to another in a fixed order, as the sciences are often supposed to do. Instead, it has to manoeuvre somewhat unpredictably to meet the varying emergencies of a changing pattern of life.

Philosophizing, in fact, is not a matter of solving one fixed set of puzzles. Instead, it involves finding the many particular ways of thinking that will be most helpful as we try to explore this constantly changing world. Because the world – including human life – does constantly change, philosophical thoughts are never final. Their aim is always to help us through the present difficulty. They do not compete with the sciences, which at present supply most of our dominant visions of reality. Instead, philosophy tries to work out the ways of thinking that will best connect these various visions – including the scientific ones – with each other and with the rest of life. So do these visions themselves need to keep changing in order to be always up to date?

2

Do ideas get out of date?

Abolishing the past

I started to wonder whether ideas get out of date some time back when I heard that, in certain universities, no philosophy was being taught except what had been published in the past twenty years. These rumours were hard to check and clearly practice varies. It seems that bumper stickers have been seen on cars in the States bearing the message 'Just Say No to History of Philosophy'. And Gilbert Harman at Princeton had a notice to that effect outside his office door. It also emerged that the term 'history of philosophy' has changed its meaning. It is now being used to describe *all* study of older writers, not just study with a historical angle. So Harman's idea is that you shouldn't read them at all and should certainly not take them seriously. Similar ideas evidently circulate at Oxford. Friends of mine at Cambridge say the situation there is less extreme, but still rather alarming. A student recently told one of them that he had spent his whole undergraduate career without reading a word of Aristotle, Descartes or Kant. At this, said my informant, 'My heart sank.'

Well, so does mine. But we need to ask just why our hearts sink, and ask too just what the people who make these changes are aiming at. Have they got a new notion of how thought works? Would they, for instance, be equally hostile to studying the history of mathematics, or indeed the history of history? Would the rise of 'Whig history' – the triumphalist view that the past has been merely a preparation for the splendid present, a view which caught on sharply after the Glorious Revolution of 1688 – seem to them a trivial matter?

Wondering about this, I remembered some things that happened in the Thatcher years (1975–90), when the recent storm of cuts first threatened British universities. Administrators, sternly told to economize, saw that the quickest way to do it was simply to abolish some subjects altogether. This would save them from awkward conflicts with more powerful empires and would harmonize with the mystique of 'centres of excellence' which were then in fashion. These centres were supposed to be big schools in which the study of a given subject would be so well covered that no other departments would be needed at all. Thus, ideally, in Britain, all the physics could be done at Manchester, all the economics at the London School of Economics, and all the philosophy (if any was still needed) at Oxford.

Down with philosophy?

This idea caught on and, since philosophy departments were usually small, universities did indeed start to close them. Eight of them in Britain went in the end. (When our department in Newcastle went in 1986 it was accompanied on the butterslide by metallurgy and all the Scandinavian languages.) As one consignment of philosophy after another went overboard, it struck me that nobody was saying that this ought not to happen. Nobody was suggesting that the subject was important in itself – that universities needed to teach it; that, if they stopped doing so, they would become, in some sense, hardly universities at all. I will try to explain what this means in the course of this book.

Fired by that thought, I wrote to a number of the eminent philosophers of the time saying, in effect, 'Do something! Write to *The Times* [which was what one did in those days]. Let people know that this is important.' But nothing much came of this. Only one person did what I suggested, and that was A.J. Ayer, whose book *Language, Truth and Logic* had actually played a great part in persuading intellectuals that philosophy was a waste of time anyway. However, two of the replies that I did get still strike me as significant. I didn't keep them because they made me so cross, but I remember perfectly well what they said.

The first one came from the very distinguished Oxford philosopher Michael Dummett, and he told me flatly that it was wrong in principle to try to preserve all these provincial academic departments. Philosophy, he said, was a serious and highly technical subject which should only be studied at its own proper level. Any less professional approaches to it were useless and might even do harm. And what Dummett meant by the proper level is clear from a well-known passage in his writings where he said that 'the proper object of philosophy' had been finally established with the rise of 'the modern logical and analytical style of philosophizing'. This object, he said, was 'the analysis of the structure of thought, [for which] the only proper method [is] the analysis of language'[1] And, not surprisingly, he thought the business of linguistic analysis had now become a highly technical pursuit – something increasingly like nuclear physics – which could only be carried on by people specially trained in it.

The second letter – also from Oxford – upset me even more because it came from a philosopher of whom I had a really high opinion, Peter Strawson, and I thought it missed the point even more dramatically. Strawson said he must refuse to support this project because I was suggesting that philosophy was to be valued because of its effect on society. This, he said, was degrading to it: philosophy should only be practised and valued for its own sake ... He didn't seem to see that my suggestion did not concern the value of philosophy itself but the practical question of *who* would now get a chance to study it. Was Oxford enough?

Aims

Both these letters raised the important question: what is the aim – the point, the proper object – of philosophizing? What are we actually trying to do? And it strikes me at once that, when Socrates talked about the great dangers that threaten unthinking human existence, he didn't actually mention the danger of unexamined thought or unexamined language. What Socrates warned us against, in the last section of the speech where he defended his own life, was an unexamined *life – anexetastos bios*. And it is surely the effort to examine our life as a whole, to make sense of it, to locate its big confusions and resolve its big conflicts, that has been the prime business of traditional philosophy. Only quite lately has a different pattern of philosophizing caught on – a pattern that is modelled closely on the physical sciences and is reverently called Research.

In the physical sciences, progress can sometimes be seen as simply accumulating a string of facts as we move on from one empirical discovery to another. This seems often to be imagined simply as a means to an end, a path to greater glory, essentially just to a string of increasingly prestigious Nobel Prizes. In this process, the obstacles that are conquered are, of course, only of passing interest. Once they have been disposed of they are irrelevant to the enquiry. That is why, for many physicists, past physical discoveries have only a mild historical interest. These people always had their eye on the *next* discovery, which accounts for their exclusive concentration on the latest journals, and also for the very revealing metaphor of the 'cutting edge' of research. How is this cutting operation envisaged? Perhaps as sculpture? As surgery? Or perhaps as butchering? In any case, the aim seems to be a destructive one – to get rid of something extraneous, a form of dirt that blocks our view of our quarry. It certainly isn't creative discovery – the finding of new ways to understand the world.

So, if you want to know just how these philosophical remedies work, it may be worth your while to consult the actual engineers who dealt with these gaps – Plato and Aristotle and Marcus Aurelius and Hobbes and Hume and Nietzsche and William James and the rest – for an explanation. And it may be worthwhile to notice that these remedies are not going to stop being needed. The gaps that called for them in the first place – the great chasms that divide human life – are not going to go away, any more than the gaps between earthly islands and continents. There will always be real causes of misunderstandings between human beings, differences of viewpoint that can easily dislocate our reasoning, such as the logical gap between reason and feeling, or between determinism and free will. Philosophical engineering will always be needed.

3

What is research?

The divided brain

This originating of new methods is plainly a different kind of thinking from the specialized concentration on fixed questions that is often thought to be the core of physical science. But in fact the two enterprises must constantly work together. Scientists – especially physicists – often need to ask philosophical questions, such as 'Is this a kind of stuff or just a force?' 'Does that count as a cause or an effect?' and the like. When this happens, science too demands that gaps should be crossed and bridges made. And it does this quite as often as it asks for existing holes to be dug still deeper.

The idea of research has, however, become so closely linked to the more specialized pattern that connection with wider contexts is often quite forgotten. Then, as often happens, the brain's left hemisphere is allowed to favour its obsessive, specialized interests at the expense of the wider framework, which is needed to explain why the whole thing matters in the first place.

It is surely alarming that, at present, this kind of narrow research is treated as the central business of universities – outclassing teaching, which is seen as merely a way of passing on the ready-made results of previous research to our successors. People seem to have forgotten that actually it is often by teaching that we learn – *docendo discimus*. So today the names of distinguished scholars often appear on notices of a particular course in which they will give only an occasional lecture. They leave the nitty-gritty of explaining things in detail to overworked graduate students, so it is presumably those students who will eventually get the deeper understanding which comes from actually explaining things. This may indeed end in their becoming the next original thinkers who come up with the next startling discovery. But that wasn't the point of the original course.

All this is alarming, because this increasing specialization continually narrows what is expected of research projects, shrinking enquiries gradually down from goldmines to rabbit holes. People occasionally ask me on what topic I am doing research, and I say that I don't do any, because I'm certainly not organizing any static mining operation of this kind. I suppose that instead I try to follow the argument (as Plato said) wherever it runs, and I may finally catch it in a territory quite far from the one where it started. In fact, arguments are altogether much more like rabbits than they are like lumps of gold. They can never be depended on to stay still.

Sciences too change direction

Now of course the sort of predictable, linear progress that is expected in the sciences does sometimes happen, and it can go on usefully for a long while. But, even in physical science, it is never the whole story. It can only work so long as there is a given pattern, a preset journey which will go reliably from A to B and so on to the end of the alphabet in the same expected course. Even in the sciences, that pattern isn't always there. Often the next important discovery is going to crop up somewhere quite different, right off to the side of the expected route. Some awkward character such as Copernicus or Einstein or Faraday or Darwin mentions a thought which suddenly calls for a quite new direction, a different way of envisaging the whole subject. Fresh questions need to be looked for, not just different answers to old ones. And this is one reason why we so much need to be aware of the history of these enquiries. The meaning that is now attached to the present-day orthodoxy got there because it belonged to the orthodoxy before last, and it can't be understood if we don't know about that.

An interesting illustration of this continuity came up in 2015 when the distinguished physicist Peter Higgs explained that the work which had led him to discover his now famous Higgs boson – a particle which apparently transforms the whole shape of physics – was so far from his official line of research that his superiors would have been quite cross if they had known that he was spending his time on it. But when these unexpected ideas do strike us, they can change the scene before us entirely. As Keats put it, when a fresh translation of *The Iliad* suddenly revealed the entire Homeric world to him:

> Then felt I like some watcher of the skies
>   When a new planet swims into his ken;
> Or like stout Cortez, when with eagle eyes
>   He gazed on the Pacific – and all his men
> Looked at each other with a wild surmise –
>   Silent, upon a peak in Darien.[1]

Shifting science

Many scientists are actually beginning to suspect that something of this kind is needed for their own topics. The point has been felt particularly strongly of late since historians and other specialists failed to predict the end of the Cold War. More recently, too, it has become striking in economics, where the accepted orthodoxies dramatically failed to predict the financial disasters of 2008 and have shown no signs since of developing to fit the facts of the times. So, since the effects of bad economics affect the whole population, biologists have offered to help by suggesting new methods, not fixed, like the former ones, to a familiar individualistic concept of Economic Man, but shifting, as other evolutionary thinking does, along with the changing units of human society. Thus, in an article in *New Scientist*, Kate Douglas insisted that this is not just a superficial matter of one speciality imitating the tricks of another but a genuine, necessary change of method. As she wrote:

> For would-be revolutionaries, it's not just a question of whether economists should do biology; *it's about viewing the world through a different lens*. It's about basing economic modelling on what biology tells us about human behaviour – and how we can channel that into creating the outcomes we desire. What is the right balance between competition and co-operation? … How do we create a more equitable form of capitalism?[2] (Emphasis mine)

She is taking it for granted that economists should be aware of the main developments in their sister subjects, such as biology. Plainly, here, as often happens, the new idea that is needed in economics comes from a different branch of learning. This is contrary to the ideal of specialization which many people now have – that these branches are unrelated life-forms which cannot interbreed; that each must study only its own range of problems. Our current emphasis on specialization makes it increasingly hard for academics to break through this artificial restriction, but they surely need to do so. It is very natural that, on first sight, economists should regard evolution as someone else's business. But if it hasn't struck them yet that the species they are talking about – *Homo sapiens* – is actually subject to widespread biological forces, then the rabbit hole that they are exploring will never lead them to gold.

4

Clashes of method

Asking not 'whether' but 'how?'

Here we surely come back to our original puzzle: why do people need to study philosophy at all? And the most obvious answer is – because it explains the relations between different ways of thinking. It maps the different paths that thought can follow. And in our journey we use so many of these different thought-paths that questions about their relations constantly give us trouble, whatever other issue we may be trying to understand. We need to grasp that one path does not necessarily displace another and make it obsolete. They can both be needed as parts of a wider, more sufficient map of the whole journey.

Orthodox scientists, however, now ignore the crucial importance of this conceptual background in a way that must startle anyone who has noticed the tangled history of past thinking about science. For instance, here is the response that

Sean Carroll got from fellow scientists when he called for a little attention to these difficulties in *New Scientist*. He wrote:

> We need to think about the right way to think about quantum mechanics if we're going to understand, for example, how space–time emerges. What look from the outside like fuzzy, philosophical questions about the nature of reality, which we can debate for years and years, will suddenly become enormously relevant. They will become sharp tools for answering deep questions about cosmology and particle physics …
>
> [Interviewer] How do your peers react when you say that your philosophical position matters when you're doing cosmology?
>
> [Carroll] A lot of them just roll their eyes. They're like, 'Really? I thought we'd got rid of that kind of stuff!' It's put me outside the mainstream, but I'm OK with that.[1]

What these respected supporters of the mainstream plainly don't understand is that the past advance of science – especially of cosmology – to its present heights has not been due primarily to a series of exciting factual discoveries but to painful and careful philosophical thinking, a method which, up till the times of Einstein and Bohr, was still quite open and explicit, but which has recently been absent-mindedly dropped out of their education. When Carroll advises his readers to 'think about the right way to think' he is appealing to that solid philosophical tradition. When they reply by evidently taking the word 'philosophical' to be more or less a synonym for 'fuzzy', they show that they don't know which way science is actually facing.

Past scientific progress has not been primarily due, as current myths often tell us, to a few dramatic and exciting experiments. Although the experiments are important, the basic progress comes from steady, careful thinking about *how to think* about the whole system in the first place. It is that thinking which shows which questions to ask, which experiments to perform, what confidence to put in them. And the operative word is *how*.

If you want this point spelt out fully in contemporary terms, Michael Brooks can tell you that you need only try asking whether you can imagine one particle exerting an influence on another hundreds of kilometres away. Actually, says Brooks, little imagination is needed. 'Such entanglement – or "spooky action at a distance" in Einstein's derisive phrase – is a consistent effect in the quantum world. So that's not the anomaly. The anomaly comes in what this tells us about our perception of space and time.'[2] To grasp this, the researchers have first had to check experimentally that the cause of the strange phenomenon actually was a quantum event. There were considerable difficulties about doing this, so that, as he adds:

> it was only last year that physicists at the University of Delft in the Netherlands devised a version of the test that finally ruled

out standard information transfer, random particle fluctuations or detector snafus as the source of the effects. 'This allows us to certify that the problem is not a feature of a particular set-up … but really *points towards a fundamental feature of nature*' [said the experimenter involved, Jean-Daniel Bancal of the Centre for Quantun Technlogies at the National University of Singapore]. And the result is clear. *In the quantum world, our understanding of space-time, and cause and effect within it, dictated by our intuitive sense of how the world works, does not apply.* There is something we're just not getting.[3] (Emphasis mine)

This is just one example of what happens when we need to change gear, when we shift from one kind of problem to another. Often we make that change unconsciously, but sometimes we can see that it is needed.

Oddities of the quantum world

In short, in a quantum context we have to think differently about things like time and space, not because our ordinary ideas of these things are wrong, but because they are context-limited: they only apply in relatively normal conditions. And, although experiments can help us in establishing that this is so, what we chiefly need is not the experiments themselves so much as the kind of thinking which shows the need for them.

This is just one example of what often happens when we need to change gear; when we shift from one kind of problem to another. Often we make that change unconsciously. Sometimes we see that something of the kind is needed, but can't locate it at all. And this may be because it is so much larger than we expect; because our difficulties have moved to an altogether different scale. This is not just a matter of an occasional exception which can be fitted in somehow; it is so frequent that it clearly indicates something graver. It warns us that our whole subject-matter – that is, the whole of life – is something vastly larger and more varied than we have been supposing.

Help from the tune of life

It is surely a piece of good luck that, quite recently, a book has come out which recognizes this startling fact and sketches in the background against which we can see it clearly. This book is called *Dance to the Tune of Life: Biological Relativity* by Denis Noble (2017). Its theme is that we need always to see the relevance of our questions to the wider contexts out of which they arise. This is not to say that limited and isolated investigations (such as the recent evolutionary enquiries wholly centred on genes) have been a waste of time. As Noble puts it:

This is emphatically not a book deriding the immense achievements of the reductionist approach. On the contrary, this approach has successfully drilled down to identify the smallest molecular components and their molecular interactions. But there is a big difference between acknowledging this great success of molecular biology and adopting it as a method, on its own, for unravelling biological complexity. Pure reductionism will not work, precisely because it does not analyse the kind of complexity organisms display.[4]

Interpretative thinking like this is, of course, not only needed for science but for everyday life too. Notable examples of it occur, for instance, in our constant puzzling about free will. We often ask: could he – or she or they – actually help what they did? And the reason why people find these questions so awkward is that they haven't yet begun to wonder *how* to connect life's two aspects – the unpredictable flow of personal experience on the one hand and the fixed, deterministic system of calculation that appears on our clocks, calendars and other machines on the other. It hasn't struck them that *neither of these systems can be wholly right or wrong.* These are simply two different techniques, two approaches, two ways of imagining and representing a single process. They are like two different languages; two partial accounts of a most complex whole, which are useful for different purposes. Their relation is much like that between seeing a flash of lightning and hearing the subsequent thunder. Lightning and thunder are not really separate items; they are two partial images of a single electric discharge. Neither is an illusion; they are only both incomplete.

Similarly, when we think of an activity in two different ways for different purposes, we fit it into two quite different conceptual schemes. For instance, if Einstein has just managed to solve a difficult problem, the most useful way to understand his solution will usually be to view it as he viewed it himself – as one possible development of his, and other people's, former thoughts on the matter. This approach puts it in the context of alternative possible solutions, so it naturally assumes a free choice between them. And indeed, theorists must make this assumption all the time when they are considering possible alternative views on a subject, whatever opinion they may profess elsewhere about 'the problem of free will'.

The alternative view, which pictures Einstein's reasoning simply as an inevitable performance by his brain cells, is not likely to be particularly useful, except in special cases where the reasoning itself is faulty. In these cases its faults may indeed be of interest to the thinker's doctor. But the problem –'Why did he think in this way?' – is then no longer a problem in psychology. Instead, it has become strictly a medical problem, one about 'What has gone wrong with his brain?' and its causes will naturally be looked for on deterministic principles. Both methods are appropriate; but appropriate to quite different enquiries.

Hemisphere trouble

The brain's structure, however, is not just an irrelevance here. There is one striking fact about brains which really does bear on these cases where incompatible truths seem to clash. That is the awkward mutual dependence of the two halves of the brain.

Rather remarkably, these two very similar brain hemispheres both play a necessary part in all our thinking, and each works to correct the other. Scientists used to think that the left hemisphere actually did most of the work, because damage to it interferes with obvious capacities such as speech. But it has gradually emerged that this damage, though obvious, always affects relatively detailed matters, while right-hemisphere injuries, by contrast, can disturb the whole balance of the person's thinking. The right hemisphere is, it seems, responsible for summing up the whole situation, and also for watching to see where things are going wrong, while the left dives into the smaller details and directs the action. The right is what makes it possible for a thrush to spot a snail from a distance, and it also watches out for predators that might interrupt the thrush's attack on it. But it is the left that comes in to direct his eye and beak as he finally deals with it.

In a great range of species, including ours, these two entities co-operate well most of the time. But the communication between them is not perfect. In particular, the left hemisphere often fails to notice more distant objects which the right may know are of the first importance. It fails, in fact, to look out for new factors and to detect approaching dangers. In human life, this produces a notorious weakness known as obsession, which can lead, in argument, both to a general over-fascination with particular details and to an even more general forgetting of the central point at issue.

This wider function of the hemispheres has only been grasped lately. But of course the effects of the resulting unevenness in argument have long been familiar. Unbalanced attention to the two halves of our thinking powers is clearly the source of many well-known intellectual faults – on the left side, faults of narrowness, obsession, prejudice, pedantry and general impenetrability; on the right side, faults of vagueness and undisciplined optimism or pessimism.

But the fact that the left brain is systematically blind to its own inadequacies, while the right is quite capable of perceiving its own faults, makes these troubles distinctly one-sided. And the academic fashions of our present age, which recommend specialization and require that all insights – however obvious – should be documented, are increasingly giving the left side an undeserved and unhealthy primacy in our official reasoning.

Besides these general tendencies, we all have our own particular chronic biases, and there are certain controversies on which the two kinds of partiality lead people to line up in a remarkably predictable manner. The debate about free will is one of these. Unshakable determinism is an absolutely typical left-hemisphere stance, a position that produces immediate satisfaction, together with lasting subterranean doubts about its obvious incompatibility with some central aspects of life. There can be similar trouble when we have to think about mental illness, where we face the difficulty of relating inner, subjective impressions to outward medical data. In both these cases, simple, extreme positions are available. On both topics, those simple positions can easily be made to look final and invulnerable. But we know they only speak to a part of ourselves. Such troubles are chronic: neither half-truth can be shot down; neither champion will go away.

If we want to get past this blockage, we have to grasp that in many situations we need to take a step back and somehow to see that what look like irreconcilable alternatives can in some sense be true together. In fact, the conflict here is always between two parts of our selves – two tempers, two incomplete world-views which need to understand each other better. Difficult though this is, people do quite often manage to do it, and if they did not the world would be in an even worse mess than it currently is.

5

Signposting problems

Are there real mysteries?

This extra space – this availability of different perspectives – is, of course, what makes it possible for us to use different units, different thought patterns, different assumptions, for so many different types of problem. About free will, for instance, some choices are best understood as fixed in advance, others as remaining open. (The question is not whether there has actually been a fixing process, but merely which assumption is most useful.) Some things, again, are best measured by triangulation, others by using an electron microscope, others in decibels, others by their reaction to very distant objects, others perhaps in dols and hedons (units of pain and pleasure). Yet others – such as the relative importance to us of various elements in life, including our various beliefs – simply call on our inbuilt powers of social response and assessment; they do not call for measurement at all. If we were not social creatures, equipped with those powers, we would not be asking those questions in the first place. And when we do ask them, those inward powers are quite equal to providing answers.

These various ways of reckoning, and the units that go with them, are not, of course, themselves items in the world. They are simply patterns for describing those items, patterns made to suit our various faculties. Each of them has been devised to describe a particular kind of fact which it fits conveniently. Thus a 'foot' was originally the length of the king's foot, and an inch was the length of his thumb-joint. There was no reason to expect terms like these to apply to things of a different kind, such as thoughts.

But when physical scientists began to see the vast importance of measurement for their own investigations and to extend its use more widely, cases like human motivation, where exact measurement and prediction are not possible, stuck out like sore thumbs and began to cause alarm. 'Free will' had, so far, been treated as a normal factual problem, and various kinds of determinism and indeterminism were devised as ways of solving it – possible paths between which we would have to choose. And until quite lately, educated people faced with this dilemma usually paid a respectful homage to the idea of determinism. Though they showed, by continuing to make their own choices, that they did not actually believe their decisions were fixed in advance, they still felt bound to accept determinism as a theory simply because of their general reverence for physical science.

During the past century, however, philosophers have provided enquirers with one more alternative: *mysterianism*. This is the view that there are some questions which our minds are simply not fitted to resolve, and that free will is one of them. In order to resolve this metaphysical puzzle Noam Chomsky adopted the name 'mystery' for these cases, apparently from a pop group called Question Mark and the Mysterians. He suggested that these unmanageable questions are not really problems at all but *mysteries*, situations in which scientists should stop saying (as they always do at present), 'We do not have the answer to this *yet*,' and should simply say instead, 'This one is beyond us.' He adds that this limitation is not surprising since the cognitive capacities of all organisms are limited, which indeed is true.

Will this do? It is surely a relief to hear the learned admitting that there are some kinds of things that they do not and cannot know. But we need to ask next, which ones are they? And why?

Free will is not, of course, an isolated case. The whole relation between mind and body – which crops up in every kind of difficult instance – has become equally puzzling ever since dualism disconnected them. The prime example considered has always been consciousness. As T.H. Huxley put it, 'how it is that anything so remarkable as a state of consciousness comes about as a result of irritating nervous tissue, is just as unaccountable as the appearance of the Djinn, when Aladdin rubbed his lamp'.[1]

Does this justify philosophers today in simply treating it as an insoluble mystery? That idea seems disturbing, first, because we clearly cannot go on making that excuse all the time, classing every question that we can't answer as a mystery, and also because, today, such a move strikes us as an unthinkable offence against the current creed that Science can explain everything. This, no doubt, is one reason why, as we shall see, scientistic thought today toys with the idea of abolishing Mind altogether.

6

What is matter?

The Huxleyan twist

When confusion reaches this point we should surely ask, why are we travelling on this road in the first place? Why has all the suspicious questioning, all the puzzlement, from Huxley's day to the present, been turned towards queries about Mind – towards doubts about the Ghost rather than about the Machine? Is it not obvious, once we begin to question the terms of mind–matter dualism at all, that there is something badly wrong with the idea of Matter?

This word 'matter' is not just a handy name for all the physical things we see around us – tables, chairs, trees, human bodies and so forth. In dualist theory it was a technical term, a name for the inert substance out of which these things are supposed to be made, mere abstract stuff. Though this word 'matter' corresponds to the Greek *hyle*, wood – which is a real material of which tables and statues can be made – no such abstract wood-stuff is ever found in the real world. It is just a rhetorical device, signifying something blank for the formative powers of Spirit to work on.

This neutral material was badly needed in a dualist universe to provide some way of accounting for the actual appearance of life and activity in the world, but it never did this effectively. In fact, dualism itself belonged to an age which had become aware of physics, but not yet of biology. It explained actions apparently performed by organisms as due simply to impact – the random banging about of items already disturbed – and it attributed any resulting order to Spirit or Mind, which alone was supposed to understand it.

That is why, when once the concept of Spirit began to fade away from scientific thought, scientists found it so hard to understand activity at all. Dualists had sternly swept away the previous belief in organic forces of love and hate, by which believers in Natural Magic had explained earthly tendencies to act, such, for instance, as the tides. They rejected these forces in order to avoid idolatry by concentrating all real power in the hands of God.

It is surely therefore rather odd that, today, people who want to show that they are scientifically up to date will often say, if asked about their beliefs, that they are 'materialists' – meaning that they think the real world is entirely made of matter. This cannot mean that it consists of the blank, inert stuff which had previously gone by that name. The people who claim this title mostly seem to mean by it merely something negative – that there are no minds or souls.

Later thinkers, of course, rejected dualism itself. But the extraordinary thing about this rejection is that it has been so one-sided. Though scientists claim no longer to believe in Ghosts in the Machine, they actually only dropped the Ghosts; the Machine remained as potent an image as ever. Because they were so used to the background notion of Matter, and because actual, literal machines were becoming so important a part of the world, it did not occur to them to change their imagery. In fact, this is one of many cases where dreams and fables have proved stronger and more lasting than official scientific beliefs.

So, when Enlightenment thinkers took God, the source of all movement, out of the realm of science, the fact that activity still went on became very puzzling; became, indeed, invested with a certain Mystery. That is why, when that devoted adherent of enlightenment scientism, T.H. Huxley, tried to understand 'how consciousness could occur as a result of irritating nervous tissue,'[1] he saw no recourse except to invoke magic by talking about Aladdin rubbing his lamp. It is also why his present-day descendants find themselves proposing a mysterian solution which is surely just as hard to reconcile with their professional beliefs.

Forget the nervous tissue

There is, however, a less dramatic escape route than Huxley's, which is often useful on these occasions. It is to say simply, 'You are trying to answer the wrong question.' Why, indeed, should anyone think that consciousness is 'caused *only* by irritating nervous tissue'? It is surely obvious that, among the causes of present consciousness, there has to have been a long evolutionary development which has shaped the whole nervous system for this very function, and a way of life which makes this kind of sensibility necessary. The effective cause, then, is not just an isolated happening in the brain. It is a whole way of living adapted to such conscious events, inside a total physique for which they make perfect evolutionary sense.

Huxley, who was such an eager propagandist for the whole idea of evolution, ought surely to have seen this possibility. And I think he might have done so but for two obstacles. First, his deeply-absorbed dualist background made him see all sensibility, and particularly human sensibility, as Spirit – something supernatural and almost godlike, much too grand to form part of any natural species inheritance. And second, his lack of interest in the lives of non-human animals blocked him from attending to the vital role that their ways of living had played in human evolution. Unlike Darwin, Huxley ignored this animal past entirely. He plainly never reflected that, in order to live, even the smallest and most primitive creature has to move about in an appropriate and sensible manner, so it absolutely needs the kind of perceptions by which it can guide its movements. In fact, consciousness is not an aristocratic privilege reserved for occasional highbrows like us. It is a basic condition of all animate life.

Whatever Huxley's own thoughts may have been on this topic, the important thing here is that his naive approach to the 'problem of consciousness' caught on and, despite its obvious confusions, it still remains bizarrely accepted today among professional philosophers. The 'hard problem of consciousness', which is now held to encapsulate the difficulty of 'explaining why any physical state is conscious rather than nonconscious',[2] is still treated as a difficulty in finding a connection between two entities which have been artificially isolated – a state of mind and a particular event in the brain. This quite obscures the fact that the conscious entity is not actually the brain-state itself but the whole living person or animal. Philosophers, obsessed with those two apparently disconnected terms, still don't approach the problem by looking at the social and evolutionary context in which the relation of the two items makes obvious sense.

Without that context, mere puzzlement about consciousness naturally backs the temptation to treat it as a mystery. Thus that eager mysterian Colin McGinn thinks this whole trouble just indicates a flaw in our evolution – it shows that there is something wrong with our brains. 'The reason we cannot make solid progress with the mind-body problem is that our human intelligence is not cut out for the job'.[3] And he intensifies this sceptical drama by referring to the brain repeatedly as *meat* ('How does mere meat turn itself into conscious awareness?'[4]).

This, however, is shockingly inaccurate. Meat comes from dead animals, not living ones, and dead animals are well known not to be conscious. Nor do the live animals that are scuttling or flitting around us in a way that shows they really are conscious, owe their origin to corpses. They owe it to previous living animals and to the particular ways of life that these animals developed for their descendants. All this misleading talk of 'meat' is not – as I think is intended – a welcome shot of realism. It is just one more gratuitous twist to the perverse insistence on locating the causes of mental events in isolated states of the brain.

7

Quantum queries

Dealing with gaps

We have seen, then, that the puzzle about free will is not an isolated difficulty which might be due to an odd or eccentric development in the brain. It is simply one of many everyday difficulties about bringing together two different ways of thinking which are appropriate to different provinces of thought, ways which have to be adjusted when the two provinces must be considered together. It is, in fact, no more eccentric than the very similar difficulty that we have in bringing together the two main aspects of human health – the physical aspect, which is appropriate for medicine, and the imaginative or sympathetic social aspect, which reflects the point of view of the patient.

When we try to use these two approaches together, we often need new and inventive thinking which will actually give us a new way of understanding the ailments we are considering. And, in this situation, we are not at all likely to suggest solving the difficulty at once by the simpler device of just saying, 'Oh well then, this is an insoluble mystery.' Nor do we simply say that the *scientific* solution must always prevail over the others. Instead, we see that the difficulty is about how to extend the neat, ordered patterns which scientists have devised for particular kinds of subject-matter to a wider territory, where other considerations must come in as well. We know that when we try to use these patterns for items of other kinds, they sometimes fit badly, so there is a clash between different ways of thinking. And this occurs on plenty of other matters besides free will and health. Indeed, for example, it notoriously occurred at the foundation of quantum mechanics.

As Tim Maudlin explains, the problem here has been that:

quantum mechanics was developed as a mathematical tool. Physicists understood how to use it for making predictions, but without an agreement or understanding about what it was telling us about the physical world … This is what Einstein was upset about, this is what Schrodinger was upset about … Quantum mechanics was merely a calculational technique that was not well understood as a physical theory. Bohr and Heisenberg tried to argue that asking for a clear physical theory was something that you shouldn't do any more … They were wrong. But the effect of it was to shut down perfectly legitimate physics questions within the physics community for about half a century. And now we're coming out of that, fortunately.[1]

In fact, a different context demanded a different sort of thinking. What happened was not so much that the questions being asked couldn't be answered as that they turned out not to be the right questions.

And things do not seem to be much changed today. Reporting a recent conference on this matter, a writer in *New Scientist* remarks, as a matter of course, that 'quantum physics is well known for being weird. The theory – and the experiments that have confirmed it – rips gaping holes in our concept of space, time and reality. Most physicists simply accept this as the way things are,'[2] though some are apparently still trying to understand it.

In short, the mysterians seem to have been operating on too small a scale. The real trouble is not that there are a few exceptional problems, such as free will, which for some unknown reason are essentially beyond us. It is that all our solutions are incomplete, for reasons that can often be understood. The area of our ignorance is enormous and, by its very nature, has no outward frontiers. The methods we have been using are proving inadequate because of difficulties which tend to arise about all problems if we pursue them beyond the more obvious first steps.

The later stages of an enquiry often call for a change of method, and finding appropriate methods is itself a troublesome new problem. We know that it can be done because we have seen it done in the past, but it always needs a new kind of thinking, and there is no method-shop that carries a reliable store of these. In fact, we all have to do our own philosophizing. But it is now beginning to be understood that in such cases we are not forced to choose one language and abandon the other. These are not problems about alternative facts in the world. They are problems about alternative possible descriptions, both of which can be used in appropriate contexts. In fact, they are problems about how to make the best use of our own imperfect faculties.

Mapping the complex world

If we are somewhat shocked at this anarchic situation and inclined to think that there ought, somewhere, to be a final, objective measurement system – a universal pattern with units to which everything will eventually conform – perhaps we should reflect that this might demand a change, not in the cosmos but in our idea of ourselves. We are not, after all, animals whose whole evolutionary history has centred on becoming able to measure the world precisely. This is just one of our occupations. In fact, the sort of precise measurement that Science envisages today is quite a new phenomenon, invented recently in our culture to suit newly developed ways of thinking, and turning out to have some unexpected limits. By contrast, the various styles of measurement that we have got used to in everyday life have grown directly out of the demands of various kinds of practice, and have been continually adjusted to suit everyday convenience.

The standardization of metric units which followed the French Revolution was an attempt to move away from this practical approach to a supposedly universal, final and infallibly rational way of measuring things. It was expected to put an end to this trouble. But, as things turned out, the demands of the various sciences for detailed calculation went far beyond metrication, and they are certainly not likely to end by converging into a new, all-purpose, universally accepted final system. Measuring is, after all, a practical job like any other, and it has to be organized in a way that suits the people who have to do it at a given time. It cannot be expected to be final and permanent.

So, if we ask what has made it possible for our new scientific questioners to make their new starts – what underlies their originality – the answer must be that it always comes from their finding some new, more appropriate way of thinking, a way that comes from their shaping the right questions for their immediate purpose. Sometimes (as here) these pioneers separate two distinct aims for the enquiry, sometimes they bring together aims which formerly seemed separate, sometimes they move the whole investigation into a larger field whose relevance had not previously been noticed. But it is important that, to achieve these feats, our innovators *must already be looking at things from more than one angle*. They don't just hop absent-mindedly from one position to another, like the rest of us. They are envisaging a larger world which contains many possibilities. And the source of these possibilities is very often an ongoing conflict. For instance, if we ask why there is a problem about free will we see that the difficulty of reconciling real choice with the regularity of nature plainly calls for a more complex map than was needed for either on its own. And drawing such maps is a central business of philosophy.

8

What is progress?

Measurement and its benchmarks

The difference that a more complex map makes to our understanding becomes plain if we look at a recent article on the topic, called 'Why isn't there more progress in philosophy?' by David Chalmers. Chalmers refuses to allow any such complexity. He says that, in order to assess philosophical development, he needs to 'articulate a measure of progress and a benchmark ... The measure of progress I will use is' (he says) 'collective convergence to the truth ... The benchmark I will use is comparison to the hard sciences.'[1] In fact, he is determined to pin down philosophy to a single function and a single map, fixing it at a single viewpoint chosen in advance.

Chalmers doesn't explain why these 'hard' sciences should be used to set a standard for philosophy, rather than enquiries nearer to its own sphere, such as the study of history, or of medicine, or literary criticism. (The metaphorical use of 'hardness' to signify virility and general grandeur cannot be an adequate explanation.) But he flatly refuses to compare it with the cognitive and social sciences, writing, 'for current purposes I do not need to take a stand on how philosophy fares relative to these.'[2] He does not explain those current purposes, nor does he tell us – what seems even more important – why he is laying so much stress on convergence itself.

This is surely odd because, on the face of things, striking advances in the sciences – ones that we would count as progress – seem to be rather more often due to divergence, to an outward move towards new approaches, started by one of those disturbing new scientific thinkers just mentioned, than to a convergence. And, as Chalmers rightly points out, there is plenty of this same divergence in philosophy too. He cites a survey made in 2009, in which professional philosophers gave widely varying answers to a number of big conceptual questions about things like free will, *a priori* knowledge, the relation between mind and body and the existence of God. As he notes, these respondents are all no doubt typical enough of our current culture, since they come 'largely from departments specializing in analytic Anglocentric philosophy in North America, Europe and Australasia.'[3] This (of course) may be partly why they produce such high scores for atheism and physicalism.

But in any case their disagreements, he says, show what a mess philosophy is in and how much progress it needs to improve it. Though we do not have a similar survey of disagreements between experts in other fields, we have (he says) enough data about them to be sure that, if we knew more, we would find much more convergence on answers to big questions there than we find in philosophy. In reply to critics who point out that the sciences too show plenty of disagreements – that physics, for instance, is radically divided on the relation between general relativity and quantum theory – he insists that 'all the same, there is in physics a large body of settled, usable, uncontroversial theory and of measurements known to be accurate within limits that have been specified. The cutting edge of philosophy, however, is pretty much the whole of it.'[4]

But obviously that does not destroy its point, which is not simply to get the question settled but to wake us up and understand it. Raymond Tallis, discussing Chalmers's article, points out that, properly understood, the question 'What is the point of philosophy if it cannot solve its Big Questions' becomes 'What's the point of being awake?' To which, he says, 'the answer is, that, if anything is an end in itself – is valuable purely for its own sake – this surely is'.[5]

More questions than answers

Chalmers, however, still regrets that, while these more factual subjects are laudably fixed and solid, philosophy remains dangerously loose and mobile. He does not remark that, in an obvious sense, it has to be, because the central business of philosophy is to deal with problems that are radically unsolved, not by chance but *by their very nature* – chronic conflicts that arise constantly in new forms between distinct ways of thinking and living. For instance, as we have seen, the difficulties of understanding the relation between mind and body, or between free will and determinism, are not factual ones, like the difficulty of finding the number of stars in a galaxy. They are conceptual difficulties about finding the best way to think about the puzzling mixture of facts that we already have – facts which come to us in constantly changing forms and need to be described in changing terms.

Philosophy, in fact, is all about *how* to think in difficult cases – how to imagine, how to visualize and conceive and describe this confusing world, which is partly visible to us, partly tangible and partly known by report, in a way that will make it more intelligible as a whole. It is a set of practical arts, skills far more like the skills involved in exploring an unknown forest than they are like the search for a single buried treasure called the Truth. And because of this it is far more concerned with the kind of questions that we should ask than with how, at any particular time, we should answer them.

Chalmers finds that many of those he questions express this difficulty by claiming that 'there is no fact of the matter', that they 'accept another option', or the like. But instead of asking these awkward customers what they are disagreeing about, he deals with them by collapsing their views towards whatever seems to be the nearest terminus. This does no doubt give him convergence – of a kind. What it obscures is the close relation between how we think and how we live.

Fitting in free will

The importance of this relation struck me lately when I got an intriguing letter from an old friend whom I knew to have an exceptionally deep understanding of horses, cows and other animals, and indeed of the whole natural scene around us. She wrote:

> I have problems with what I think you are saying about Free Will in your book *Are You an Illusion?* probably because I don't understand the philosophical background. From a biological and behavioural point of view, since most behaviour and decision-making (in mammals at least) is enormously affected by lifetime experiences and innate tendencies, but the 'if, where, when, and how' these are performed are the result of life experiences; thus behaviour and decision-making does not seem to add up to what I understand as Free Will?

Of course she is right. Talk of 'free will' does involve unpredictable choice: choice seen from the position of the chooser. But when we are talking strictly *from a behavioural point of view* – that is, with our attention fixed on finding particular physical causes, which behaviourists assure us are the only real causes in the lives of animals – we can only look for them among facts about our own previous movements and those of people around us.

But then, except for deliberately restricted academic purposes, we never do think about human conduct in this way, nor indeed about the conduct of the animals that form part of our lives. When we wonder, as we often do, just what actually led someone to speak and act as they did, the last thing likely to occur to us as a cause is a passing event of this kind. We know that, instead, we need to attend to the wider context – to the speaker's whole train of thought and action and indeed to the existing disputes in which they are engaged. This is the lasting context that has led them to speak and act as they did. And of course it centres on their own motives – the state of their will.

This central importance of motive in communication does not need to be explained, because everybody who attends at all to speech – or to any other kind of utterance – knows about it already. It is not the kind of technical 'philosophical background' that needs to be mentioned. It can be taken for granted, as much when we ask about the reactions of horses or cows to our communications as for human beings. And the motivation

that is typical of any species is as central a part of that species'
biology as its digestion or its manner of walking. Thus, *from a
strictly biological point of view*, motivation is always relevant. The
behaviourists' attempt to exclude it, along with everything else
subjective, is just idle theorizing.

How, then, does this background bear on the prospect of
convergence?

No doubt it is true that we do want thinking humans to
converge eventually – to agree in their views about these large
issues. But we only want this if it will indeed give us a larger
truth than the one we have been discussing. And we certainly
don't have any reason to expect each event to be accounted for
completely by a single cause. So, if truth of some kind is our
aim, it must surely be this larger, more distant truth, not a simple
convergence on a nearer one.

But to look for this larger truth would mean reformulating
the whole question that we are trying to answer, which would
interfere with the facts that are Chalmers' main interest – facts
about the answers philosophers are giving to 'big questions' as
now conceived. He names these big questions in everyday terms,
such as 'How do we know about the external world? Is there a
god? Do we have free will?' etc. But, so as to get clear facts about
the current state of philosophical progress he formulates each of
these strictly on a yes-or-no model, collapsing answers which
insist on ambiguity in one direction or the other. He thus gets

static answers which quite obscure the evidence for intellectual change that he is supposedly looking for.

For instance, adding or removing the idea of God is not just changing an empirical detail, like adding or removing Australia from the map of the world. It is much more like changing the idea of that world as a whole. It alters the whole subject-matter. For, unless we have signed up as dualists, it is natural to us to think of this world as an inhabited one, and the question about what beings inhabit it is central. It is not possible to imagine that world, as the Vienna Circle recommended, as consisting only of physical details. In actual life, each of us has a world with a great background which our culture makes ready for us, including a whole population of human and non-human creatures, forces, atmospheres, opportunities, customs, tendencies, ideals, dangers and challenges. As Iris Murdoch has sharply pointed out, this 'culture' is not just a matter of a few recent films and fashions; it contains everything that we believe in, including our fashionable views about science itself:

It is totally misleading ... to speak of two cultures, one literary-humane and the other scientific, as if they were of equal status. There is only one culture, of which science, so interesting and so dangerous, is now an important part. But the most essential and fundamental aspect of culture is the study of literature, since this is an education in how to picture and understand human situations. We are men and we are moral agents before we are scientists, and the place of science in human life must be discussed in *words*.[6]

So, the words in which scientific conclusions are expressed – words like evolution, gene, autism, information, multiverse, learning difficulties – along with the materialistic metaphysic that is believed to underlie them, can have as powerful an influence on our health and sanity as does our diet or our clothing. And those words are sometimes so habit-forming that it is really hard for the philosophical positions behind them to be changed.

9

Perspectives and paradoxes: Rousseau and his intellectual explosives

Philosophical positions, however, do actually change. So how do the skills that produce these changes really work? What is happening when Darwin or Einstein or Copernicus, or indeed Aristotle or Descartes, take us to a new viewpoint?

Historians sometimes treat these achievements either as something inevitable or as a kind of miracle due to individual genius. (This is why some misguided people demand a further dissection of Einstein's brain, as if that would explain his discoveries.) But what is really happening is something both more obvious and more interesting than these personal shifts. It's a matter of perspective. These original thinkers have stood right back from their local problem. They have looked at it in its wider context and seen how it connects with wider questions. They have been using telescopes as well as microscopes, so as to take in a larger subject-matter. In short, they have been philosophizing. Similarly, the people who have decided to remove God from their world-map are not just making a geographical change, like removing Australia. They are shifting the standpoint – the perspective – from which this view is taken. They have decided to move inwards and to cut out the wider spiritual context which had, till now, been thought relevant. They have decided to study a smaller world.

Now this business of looking at life as a whole – finding suitably wide contexts to give sense to our immediate problems – is philosophy's distinctive activity. It is what makes it an occupation that matters to all of us. It is not just one speciality among others that somebody might take up. It's a kind of conceptual geography which fills in our background. Indeed, it works like the basic grasp of earthly geography which lets us see why the many different maps of the world – political, physical, barometrical and so forth – that appear at the beginning of our atlases are all necessary, but are all different. Philosophy looks at the positions of our various ways of thinking and tries to map their relation. It's a way of making sense of the whole. Now the religious parts of our existing maps – the parts which deal with matters like gods – are notoriously highly various and may often be misleading. But that does not show that the areas they map are non-existent.

Thus the reason why some philosophers are eventually remembered is not that they have revealed new facts, but that they have suggested new ways of thinking which call for different ways of living – paths which were really needed. Repeatedly, they have brought absurdities to the attention of their age by graphic contrasts, often by displaying current customs against a new background and pointing out the strange assumptions that are distorting them. After this, thought can go on in new ways, new directions which do not have to put their predecessors entirely out of date. There may well be room for both of them.

For instance, when Rousseau started his book *On the Social Contract* by writing, 'Man is born free, and everywhere he is in chains,'[1] he was lighting up some crashing discrepancies between theory and practice which had to be investigated if current problems were ever to be properly dealt with – discrepancies which still trouble us today. Similarly, when this same Rousseau pointed out the strangely unnatural way in which babies were being reared – babies who were removed from their mothers, bandaged onto boards and handed over to nurses who might well not care much about them – people started to notice anomalies in their whole idea of what 'nature' is, and how it relates to our species. These anomalies had never struck them before. More immediately, they also started for the first time to pay some serious attention to the detailed facts about small children, as they have gone on doing ever since. And since we humans all start our lives as babies, this is a pretty relevant matter to enquire into.

10

Mill and the different kinds of freedom

In the first chapter of *On Liberty*, John Stuart Mill wrote:

> In the eighteenth century, when … what is called civilization and the marvels of modern science, literature and philosophy [were all the fashion], with what a salutary shock did the paradoxes of Rousseau explode like bombshells in the midst, dislocating the compact mass of one-sided opinion and forcing its elements to recombine in a better form and with additional ingredients![1]

In short, something negative – our simply *not* thinking about these obvious problems – has been a major cause of our present troubles. Mill points out that explosive interventions like these are salutary even if they introduce their own faults and don't correct previous mistakes completely. The searchlight of strong attention that they set up has a lasting effect in making it hard to ignore the issue thereafter. It is indeed interesting that our forefathers apparently could not see through their previous muddled ways of thinking until someone like Rousseau lit them up. But then, what are we taking for granted today that will be seen through tomorrow? The assumptions that produced those earlier customs were strong. They simply persisted till an appropriate shock shifted them – till they were plainly stated in forms that could not afterwards be accepted. That strong imaginative impact was enough.

And if we ask whether these outward moves should be counted as *progress*, the answer is that they surely should, if that just means that they are what needed to happen next. Often, indeed, these new moves are necessary to prevent existing doctrines from actually contracting, as they would if these basic matters were still ignored. As Mill says, in calling for free discussion of these topics:

> [Without this discussion] not only the grounds of the opinion are forgotten but too often the meaning of the opinion itself. *The words which convey it cease to suggest ideas* ... Instead of a vivid conception and a living belief, there remain only a few phrases retained by rote; or, if any part, the shell and husk only of the meaning is retained, the finer essence being lost. The great chapter in human history which this fact occupies and fills cannot be too earnestly studied.[2] (Emphasis mine)

Which freedom?

Mill may well have been thinking here chiefly of religious creeds. But today's political and economic clichés, and indeed today's philosophical clichés, can fill the role just as well. A remarkable example of the ambiguities that collect here can be seen if we notice the various political appeals that are now continually made to the idea of freedom. This idea can act as a simple negative, standing equally for the removal of almost anything, as in the phrase 'freedom from want'. But we always need to ask, freedom from what? for what? and for whom? At present, its main political application tends to be 'freedom for the market', meaning the removal of restrictions on trade and finance – a removal which provides those in charge with freedom of exploitation, that is, the chance for somebody to make more money out of other people.

This of course has the exact opposite effect from the previously popular idea of 'freedom for the workers from oppressive conditions of labour'. And the symbolism now favoured for this commercial freedom – the imagery of 'the nanny state' – is designed to discredit that earlier idea. Appeals for market freedom tend, of course, to call for longer working hours, lower wages and less protection of labour – in fact, for the withdrawal of those advantages which the working population did gradually manage to gain during the last two centuries. But, while this greater fairness does, of course, have its own appeal, people in control of money find that, imaginatively, it is a less attractive image today than simply being 'free'.

All this just gives us one more example of the subtle way in which our thoughts depend on a mass of unstated assumptions and images, very much as our physical life depends on the hidden shifting masses of the earth beneath us, of which we know nothing. We don't notice this background till things start to go wrong – until, so to speak, the smell coming up from below is so bad that we are forced to take up the floorboards and do something about it. This is why I have often suggested that philosophy is best understood as a form of plumbing. It's the way in which we service the deep infrastructure of our lives – the patterns that are taken for granted because they have not really been questioned. This critical activity is something both deeper and more outward-looking than just mapping the structure of contemporary thought and language, which sages such as Michael Dummett seem to have thought was the whole duty of philosophers.

11

Making sense of toleration

Another useful piece of plumbing was done in the late seventeenth century, when John Locke and others discovered the concept of Toleration. During most of that century people throughout Europe had assumed that they must not endure disagreement about religion. If they couldn't agree on a single truth about it, they must just go on fighting till they did, and meanwhile individual heretics must all be converted or punished. The idea that different opinions could perfectly well be allowed to exist side by side was seen as culpable weakness, leading to anarchy. What eventually struck Locke, and what he managed to express in his writings, was that this system of competing dogmas can't work because the truth is simply too complex to be expressed in any single formula. Nobody ever has the whole truth, and people who grasp different parts of it can, in fact, perfectly well live peacefully together. Indeed, that may be the best way of putting the partial truths together in the end.

This 'discovery' was not, of course (as scientific discoveries sometimes are), simply a matter of finding a new ready-made fact, such as that the earth goes round the sun. It did involve noticing some facts about the world, notably the ways in which people actually react to restraint and to contradiction. But it also called on people to change their reaction to these challenges – to be prepared to understand them in a new way. This is not like simply facing in a different direction. It is much more like inventing a new musical instrument and working out how to play it.

Locke and the people who worked with him had to learn *how* to tolerate what had previously seemed intolerable, which includes learning how to do business with people they had previously thought were outside the pale. They had to learn, too, how to look at the outer borders of this toleration and decide what must still be regarded as intolerable.

In fact, toleration, like all big philosophical ideas, is a very complex instrument, as hard to play as the cello or bassoon. This is why we still have so much difficulty in learning how to handle it properly and why we still need to go on thinking out the ideas behind it. And the other ideals round which we try to structure our lives – ideals such as justice, equality, freedom, compassion, fraternity or sisterhood – all involve puzzling conflicts. They are as complicated as they are attractive. Yet they all have to be thought out and used together by the whole orchestra.

These ideals are, of course, ones that have been central to the message of the Enlightenment, a message which we now assume is the obvious framework for any decent human life. But this Enlightenment message itself wasn't always obvious. It didn't drop ready-made out of a machine called History. It had to be worked out, devised with a great deal of hard, grinding labour by philosophers like Locke and Rousseau, and it still has to be thought through with increasing effort up to the present day and beyond. In every age, more work of this kind is needed because the truth about the world is as complicated and as fast-changing as the world itself. But this doesn't mean that earlier ideas about it have grown obsolete. It means that the gaps in and around those ideas still give trouble. They simply have to be thought through now even more carefully.

PART TWO

TEMPTING VISIONS OF SCIENCE

12

The force of world-pictures

The exploring eye

Are we getting any clearer about what is the real aim of philosophical enquiry?

One thing that is already clear surely is that it can't be at all like the aim of any particular science, such as crystallography or ornithology. Physical sciences like these spiral inward and down towards particular parts of the truth, parts which sometimes *are* ready-made facts. That is why this pattern is represented as somewhat like mining by the current notion of research. But is there then something wider which forms the aim of Science as a whole?

That aim would have to be something very general – the sum of all the particular truths, along with the relations between them.

But here it overlaps with the aim of Philosophy itself, whose primary business is surely with these relations between different kinds of thinking. Philosophy therefore ranges indefinitely outward, always looking for helpful new connections – new patterns of thinking and living that explain the ways in which we think and live now, while the particular sciences each concentrate on their own peculiar province.

So it is quite proper for nuclear physicists to be specialists, knowing more and more about less and less. Philosophers however face the other way. They are trying to do almost the opposite – to find patterns that will restructure the whole range of experience so that we can live differently. They work to extend our repertoire. They can bring landscapes in sight that nobody even knew existed before. We may call this 'progress', but, if so, we certainly don't mean that it is just the next step on the same staircase.

Of course there is not a complete division of labour here between science and philosophy. As we have seen, physical scientists do sometimes have to widen their views so as to shift their focus, and philosophers too must sometimes deal with detailed technical questions. But in their general balance these two approaches really are opposed – not because they are at war, but because they serve quite different needs. Nuclear physicists are normally addressing a limited audience of specialists –people who already share much of their knowledge and want to develop particular aspects of it.

But the philosophers' business is not – as some people mistakenly think – merely to look inward. It is to organize what concerns everybody. Philosophy aims to bring together those aspects of life that have not yet been properly connected so as to make a more coherent, more workable world-picture. And that coherent world-picture is not a private luxury. It's something we all need for our lives.

Thus world-pictures – perspectives, imaginative visions of how the whole world is – are the necessary background of all our living. They are likely to be much more important to us, much more influential than our factual knowledge. This is obvious when we notice the way in which bad and misleading world-pictures can persist despite plainly contradicting the facts – for instance, in the case of climate sceptics, whose traditional world-views remain unchanged whatever new evidence comes up to disprove them.

Shapes and structures

We all have these background pictures, and we usually get them automatically from the people around us. We often don't ask where they came from in the first place. But, if we do ask about that, we shall probably find that they have been shaped by earlier philosophers who have influenced our tradition. Very often their details emerge from earlier clashes between existing authorities, as, for instance, the warfare between Catholics and Protestants emerged from rival interpretations of the Bible and of sages such as St Paul. And it is interesting to notice how a similar thing happened later, within the church of Russian Communism. Soon after the Revolution of 1917, rival interpretations of Marxism, which were already simmering in the background, became vocal and claimed ultimate authority. This led to savage heresy-hunting by both sides, in which sins such as 'deviationism' soon became offences that could cost people their lives.

This example shows usefully how the source of religious feuds often lies, not in a religion itself, but in the political and imaginative background that has dramatized it – that is, in the accompanying world-picture. In these influential debates, the tribal fear and resentment that underlies so many apparently harmless divisions finds its function and justifies itself. Tribal names such as 'Prods', 'Romanists', 'puritans', 'infidels' or 'god-botherers' can work on their own to convey hostility, fear and disgust, without any clear idea of what the division means. So this is the point at which the work which (as we have seen) is characteristic of philosophy – the work of mutual understanding that can lead to reconciliation – becomes necessary and finds its place. When once the political and imaginative source of the feud vanishes – as it now has in Russia – it becomes clear that the matter supposedly in dispute was trivial.

And even before that political shift, astute philosophers can, if they shout loud and appropriately, show the need for a change. This indeed was what Rousseau did, and so did his many fellow prophets of the Enlightenment – Locke, Descartes, Hobbes, Hume, Kant, Mill, Marx and Nietzsche. They all shifted the current issues of their times by shouting in a way that still influences us today. And so, before there was any question of the Enlightenment, did earlier thinkers such as Plato and the other Greeks.

Persistent problems

Thus philosophy doesn't usually get obsolete, any more than rivers or mountains simply vanish on their own. Thought goes on being needed because the difficulties in life that called for it in the first place are still there. Thus, ever since the mid-eighteenth century reformers have repeatedly asked for more sexual freedom, and of course they gave good reasons for it. They usually viewed this issue as simple because they didn't consider the price that would have to be paid for it. Shelley put the case eloquently enough in *Epipsychidion*:

> I never was attached to that great sect
> Whose doctrine is that each one should select
> Out of the crowd a mistress or a friend,

> And all the rest, though fair and wise, commend
> To cold oblivion …
>   … the beaten road
> Which those poor slaves with weary footsteps tread
> Who travel to their home among the dead
> By the broad highway of the world, and so
> With one sad friend, perhaps a jealous foe,
> The dreariest and the longest journey go …

These were the kind of thoughts that led him to go off with Mary Godwin, leaving his wife alone and helpless with her son and daughter. She ended up in the river, and Shelley was startled to find that he had lost custody of his children.

This sort of disastrous confusion went on more or less unchecked till the mid-twentieth century, when changing customs gradually forced governments to give more attention to the matter. Since then, administrators have tried to arrange things so as to balance the various claims involved more fairly. They have managed to limit many of the more obvious injustices, but the resulting systems are still notoriously imperfect.

People quite naturally make conflicting demands on each other, and customs naturally grow up in a way that sometimes embitters these conflicts. Those conflicts have shaped the way we live and think. They have deep roots in the soil of our lives and they go on developing there in their own characteristic patterns

until somebody comes along and rethinks them drastically.
That, indeed, is the reason why people who refuse to think for
themselves philosophically so often end up trapped in bits of
older philosophies that they have unconsciously taken on board
from their predecessors. There is no thought-vacuum into which
these people could escape from their whole tradition. We just
have to start from where we are.

The alternative to being controlled by past thought in this
way is to attend directly to the world and to what these earlier
philosophers have actually said about it. We can then see how
it relates to our life today. If we do this, we shall often find
that these people's message was far more subtle than the crude
versions of it that are still working in the tradition. (This is
particularly obvious today in the case of Darwin.) In fact, earlier
thought still throws out shoots that can help us today. The reason
why these philosophers caught the attention of their times was
(as we have seen) not so much that they had solved particular
problems as that they had lit up life from unexpected angles.
They suggested not just new ideas, but new concepts, whole new
ways of thinking and living.

Of course, none of these new approaches solves all our
problems. But each of them gives us a fresh stance, fresh tools
for the endless balancing act by which we try to understand our
confusing world. We can see how influential these suggestions
still are, not just because people today still draw their favoured

quotations from (say) Marx or Nietzsche or Plato or Buddha or Darwin – but because current thinking as a whole is still often visibly shaped by these people; coloured through in a way that the people using it are no longer aware of.

13

The past does not die

So, how could it be plausible to think that all these ideas are out of date and we can now forget about them? How could it not be necessary for us to attend to these still influential obstacles affecting our lives? The point is not just that – as I've suggested earlier – we need to check the details of past philosophies to protect ourselves against distorted versions of these people's message that are still working in our tradition. We need also to attend to these mighty mountains and rivers, these treacherous marshes and poison-wells and volcanoes in the landscape for their own sake. We need to understand them because they have shaped the whole pattern of life that we still live by. They are still active features of our present life, parts of the tangled landscape through which we have still to travel. In fact, the reason why we need to learn about the history of philosophy is just the same as the reason why we need to learn about the rest of our history:

namely that, without grasping that past, we can't hope to deal with the present.

On the political scene this is obvious. We understand, for instance, that, if we haven't grasped the history of the ravenous way in which Western nations competed to gobble up other countries during the nineteenth century, we can't hope to know why so many people in those gobbled countries still feel bitterly resentful towards ourselves. Dangerously ambiguous terms like 'Crusade' still constantly remind us of these troubles. And we know that the Chinese haven't forgotten the Opium Wars. Yet we are continually surprised to find that historical epochs don't just succeed one another randomly like successive spinnings of a roulette wheel. They are phases in a continuum, an organically connected development. Often, therefore, you really cannot understand where you are now without grasping the background that tells you how you got there.

And if this background is necessary for understanding politics it is still more necessary for our moral and intellectual life. Without it, we can't really make sense of current conflicts. In particular, any student who is now expected to study the philosophy of the past twenty years without being told about the long sweep of history that produced it is surely doomed to frustration. And these students have all the more right to be cross about this because (as we have seen) this misdirection affects not just their knowledge, but their whole world-view, their imaginative understanding of life. We need to grasp the story of our past intellectual evolution so as to understand where we are today, just as much as we need to know about our past biological evolution.

Philosophy, in fact, is not just one specialized subject among many, something which you need only study if you mean to do research on it. Instead, it is something we are all doing all the time, a continuous, necessary background activity which is likely to go badly if we don't attend to it. In this way, it is perhaps more like driving a car or using money than it is like nuclear physics. And perhaps it is more like music than it is like any of these things, because its effect is primarily on our imagination. It is a potent background; it profoundly shapes our inner life. And, like good music, good philosophy does not easily get out of date.

Imagination, belief and warfare

So does this persistent philosophy, then, supply us with our beliefs?

It affects our beliefs because philosophizing well or badly affects our attitudes, which naturally shape our opinions, and therefore our behaviour. Indeed, most societies centre their lives round certain characteristic opinions or beliefs. Some societies take those beliefs literally and express them directly in action, as, for instance, the Aztecs centred their lives round belief in the importance of rituals of human sacrifice. But for others their beliefs have rather an imaginative or symbolic importance.

Thus, if we ask about Christian attitudes to war and conquest, we find a situation that is far from simple. On the one hand, Jesus Christ himself said plainly that his kingdom was not of this world, and forbade his followers to fight in his defence. But when Christian churches began to get political power, the people leading them were involved in conflicts with opponents whom they saw as God's enemies, and they were often quite willing to engage in literal war against them.

What made this shift between symbols and real life easier was the extraordinary power of the human imagination. Lively symbolism had already envisaged inner conflict as a form of warfare. Thus St Paul readily encouraged people to 'fight the good fight of faith'[1] and so forth. And when Constantine adopted Christianity as the official religion of the Roman Empire, the military notion of order which had always ruled Roman life latched on quite readily to the new ideals, using language which shifted constantly between a symbolic and a literal meaning. These shifts have been particularly marked in the poetic language of hymns and psalms, where they are still active today. Thus, peaceable nineteenth-century citizens were not at all surprised to find themselves singing hymns like 'Onward Christian Soldiers' or, even more excitingly:

The Son of God goes forth to war

    A kingly crown to gain,

His blood-red banner streams afar,

    Who follows in his train?[2]

People knew, of course, that this language was metaphorical. But in the turbid waters of our imagination these different levels of meaning don't remain distinct. It's an important part of my theme in this book to notice how far the dreams, the visions, the myths that surround our official beliefs can carry us into areas that we never anticipated from our starting-point. And when (for instance) actual wars are going on in the world, the application of these images can become even more confusing.

Thus my father, who was sent out to the front in France as a young army chaplain in 1916, was appalled to find that the men dying there expected him to tell them just what it was that they were dying for. That grim experience altered his views about the whole business of war, and a great deal else, for the rest of his life. He could not simply translate the symbolic language of peace-time to cover this appalling reality, as chaplains were apparently expected to do. Instead, open opposition to war became for him a chief business of Christianity, and indeed central to his political life. Of course the wider culture in which he lived did not validate this divorce between symbol and reality. But conflicts surrounding this kind of revelation – this clash between symbol and experience – became, and have remained, a crucial part of that culture.

14

Scientism: The new sedative

What I have said so far is, of course, a familiar history. We know about the shift between fantasy and reality, and we know that this shift is not just hypocrisy, but a part of the normal working of our imaginations. What I want to say next, however, is less familiar. I want to point out that the idea of physical science has itself a strong and effective symbolism. Many years ago now, this idea of science succeeded to the position of authority in our culture which used to be held by religious creeds. It is now the central example of a compulsory doctrine, something that has to be believed. It is an oracle. And this status has disturbing effects on the way in which we now think about science itself.

Instead of seeing the physical sciences as real, but limited, sources of knowledge about material facts, we are now called on to revere them as the metaphysical source of all our knowledge. Science fiction helps this confusion and can add great emotional force to its products. Physics and chemistry do not then appear just as two stars among many but as parts of a kind of super-sun, the final form of knowledge, the destined end-product of all learning, a terminus for which all other kinds of thought are just provisional sketches.

This is the vision of science now on offer; the creed of scientism. If you doubt that such a creed exists, let's look at a recent manifesto for it, issued by Professor Laurence Krauss, cosmologist and professor of physics at Arizona State University: First, Krauss got rid of philosophy:

1. All questions of philosophy are either meaningless or can be answered by Science.

2. Philosophers are of no use to scientists.

And by getting rid of philosophy he has, of course, got rid of any system that might show science as just one way of thinking among others, since all such systems are philosophical. Next, he explains *why* science reigns supreme:

3. Science has authority because it is based on physical evidence – scientific claims will therefore always overrule philosophical claims.

And finally he sums up the ruling position which it has now reached:

4. Science provides *the ultimate account of the basis of reality – the ultimate metaphysics –* but it substantively

changes the questions, getting to the correct ones rather than the meaningless philosophers' ones.[1] (Emphasis mine)

The magazine *New Scientist* expressed this scientistic creed clearly in the headline with which it commented on the discovery of the Higgs boson, which had just been acclaimed as the ultimate particle. But *New Scientist*'s comment on its front cover was 'Forget the Higgs. Now we're searching for *the root of reality*' (emphasis mine).[2]

Krauss's first two claims here seem to clash rather sharply with his last one. Science (it seems) rules that philosophy is not needed, yet it can apparently now start philosophizing on its own ('science provides the ultimate account of the basis of reality – the ultimate metaphysics'). We might ask how it can do this when it is also claiming (3) that its authority is always based on physical evidence? 'Ultimate accounts of the basis of reality' would surely have to apply to reality as a whole. They couldn't rest on particular physical facts.

In the early twentieth century scientism was needed because the soothing, reassuring beliefs which – despite religious wars – had kept all educated Europe in balance for a dozen centuries were being seriously undermined. Many people no longer found it possible to rely on a hidden, assumed Christianity to supply their lives with a happy ending. Something else was needed instead.

Cognition and circuitry

So what were the ideas behind scientism? As so often happens, we need to look directly at the background feud, the ongoing dispute that has brought this matter to the surface. That dispute was already raging in the nineteenth century and it actually concerns prestige. It was about whether the real world is essentially made of mind or matter – which means, essentially, 'which is the top discipline, science or some part of the humanities?'

This question certainly can't be settled by any particular physical facts, nor of course by mental ones. It is a philosophical question, so it has to be dealt with, like all such questions, by getting a better understanding of what is at issue. For this we have to ask first, 'Why are these two things, mind and matter, being opposed and set in competition in the first place?'

This is not just an enquiry about what the objects around us are 'made of' in the everyday sense. That would be a real physical question, quite properly answered by physical science. But if that science provides universal answers, it can be expected to make sense of human life as a whole, which must include dealing with every sort of question. What, then, can it tell us about, for instance, the causes of warfare, depression, sport, obsessiveness, the meaning of sex, mortality, etc.? And here physical science is of no immediate help to us. Usually, it simply isn't relevant.

Scientism, however, insisted that these physical sciences *will* become relevant quite shortly – in fact, they will give satisfactory answers once they have completed their current research programmes. Thus E.O. Wilson, in the rousing conclusion of his monumental tome *Sociobiology*, explained how the sociology of the future would be able to deal scientifically with topics which now seem to involve human thoughts and feelings, because it will reduce them to their underlying brain mechanisms. This new enlightenment, said Wilson:

> must await a full, neuronal explanation of the human brain. Only when the machinery can be torn down on paper at the level of the cell and put together again will the properties of emotion and ethical judgment come clear … Stress will be evaluated in terms of the underlying perturbations and their relaxation times. Cognition will be translated into circuitry. Learning and creativeness will be defined as the alteration of specific portions of the cognitive machinery regulated by input from the emotive centres. Having cannibalized psychology, the new neurobiology will yield an enduring set of first principles for sociology.[3]

Wilson (who, incidentally, no longer talks like this today) made that generous offer in 1975, which was actually something of a high-tide moment for scientism, as it was also for science fiction. At that time, scientistic claims like this were heard on all sides, and they naturally produced the obvious response, 'OK, go ahead and do it.' An attentive audience awaited results, but these haven't been very encouraging. It soon became clear that you can't actually evaluate stress simply in terms of the underlying perturbations and their reaction times, because *evaluating* is a distinct kind of job, calling for a quite different conceptual framework. Nor can we settle all the questions that arise about learning and creativeness simply by defining these things as 'alterations in the cognitive machinery'. Brain research has indeed intensified and is increasingly busy today. But it still hasn't explained how to relate the realms of mind and matter.

In fact, this whole reductive programme – this mindless materialism, this belief in something called 'matter' as the answer to all questions – is not really science at all. It is, and has always been, just an image, a myth, a vision, an enormous act of faith. As Karl Popper said, it is 'promissory materialism', an offer of future explanations based on boundless confidence in physical methods of enquiry. It is a quite general belief in 'matter', which is conceived in a new way as able to answer all possible questions. And that belief has flowed much more from the past glories of science than from any suitability for the job in hand. In reality, not all questions are physical questions or can be usefully fitted to physical answers.

We will look further into this concept of materialism, and its younger sibling, physicalism, in Chapter 20.

Why two worlds?

But, before doing that, we need first to look more carefully at the problem of dualism – to ask what need there was to divide the world into mind and matter in the first place. Dualism proceeds as if these were simply two distinct materials – perhaps like iron and nylon? – so that we would only have to decide which of them is the more fundamental.

But that conception of a twofold world is an outdated muddle. It arose in the sixteenth century because only two sophisticated ways of thinking about the world were then available to enquirers – essentially, the new physics on the one hand, and theology, with its doctrines about the fate of the human soul, on the other. Because there was no obvious way of relating these two, people felt forced to choose between them. Descartes therefore decided (and, if he hadn't, somebody else would have done) that theological and psychological thinking applied only to the world of the mind, while physics dealt with the separate world of matter. The only link between the two was that both were the work of God. As Newton put it, Nature and Revelation were God's two books which he had given us to study. Both were important but – since God was the author of both – at the deepest level, theology necessarily trumped physics. And for a century or two most people, including most scientists, were quite satisfied to accept this.

Nature and its many forms

Since that time, however, the scene has been transformed. First, the physical sciences themselves have branched out into a galaxy of different methods entirely unlike the highly abstract physics of renaissance times – methods suited for particular different kinds of subject-matter. And a whole set of other galaxies have developed around them to deal with other kinds of topic. Historical methods have grown up for dealing with the past; linguistic methods for dealing with language, and plenty of meta-methods – philosophical methods – for dealing with the relations between all these different kinds of enquiry. Perhaps most disturbingly, too, the social sciences have arisen to deal with matters nearer home – ways of asking about the inner complexities of human life.

In short, it has become clear that *there aren't two separate kinds of stuff at all.* 'Mind' is what we see if we look at the world from one point of view – through one kind of lens, asking one kind of question – and 'matter' is what we see if we look at it through a quite different lens, asking different questions. The gap between these two is more like a change of lighting than a shift of substance. One view centres on subjects, the other on objects. One shows the inner world, the other the outer one. But these are not actually two separate worlds. They are aspects of the same one – the single, all-purpose world that belongs to all of us. And, since they are both genuine aspects of that one world – both essential parts of our lives – both are equally real. There is no kind of reason to treat that somewhat mysterious metaphysical entity 'matter' as something real, something solid, and 'mind' as something fluffier, less genuine.

Can there be a social science?

The social sciences, meanwhile, have taken on the difficult work of explaining how this many-sided, many-structured world works. And we know that, though they sometimes get things wrong, their work is necessary. But our scientistic prophets find this whole suggestion – that there can be sciences which do not reduce to physics – deeply shocking. Thus, Lewis Wolpert shows a real horror at the whole idea of sociology. He writes:

> In a sense all science aspires to be like physics, and physics itself aspires to be like mathematics … In spite of recent successes, biology has a long way to go when measured against physics or chemistry. But sociology? What hope is there for sociology acquiring a physics-like lustre?[4]

This plaintive cry of physics-envy shows just how distant the dreams of scientism have become from scientific reality. Wolpert is himself a distinguished embryologist, so he knows perfectly well that his own science could not possibly be reshaped to resemble physics. And it does not need to be. Embryology is about embryos. It is not designed, like physics, to deal with every kind of physical matter. So it uses forms that suit its particular topic. And so does every other special branch of science. These branches don't have to be unified or reduced to something else; they can all just be used together in parallel.

But still that dream persists of a single ideal 'scientific method', a universal super-study that could really explain everything. And Wolpert's notion that 'in a sense all science aspires to be like physics' shows how physics itself has now been elected to that sublime position. He is quoting Walter Pater's remark that 'all the arts aspire to the condition of music', which only meant that they must not be distracted from their central aim, as poetry can, by the oddities of particular subject-matters. But the idea that 'physics itself aspires to be like mathematics' suggests something much more extreme: it suggests that sciences ought to have no subject-matter at all and that physics can eventually live without one.

This, however, is a really strange dream, myth or fantasy. Physics isn't, like logic or mathematics, a purely formal study. It gives us real information about the world. But it never pretends to tell us the whole truth about that or anything else. Like every particular science, physics too is a one-sided story, an abstraction, a view seen through a single window, a set of photos taken from one particular viewpoint. In order to use it, we always need to put it together with a background from the other sciences, and with a great mass of informal knowledge which counts as common sense. It could never be the ruling centre of our thoughts. In fact, physics itself never makes Krauss's wild claim to supply 'the ultimate metaphysic – the ultimate account of the basis of reality'.

PART THREE

MINDLESSNESS AND MACHINE-WORSHIP

15

The power-struggle

The blank within

So why should anyone ever have expected a physical science to supply the ultimate account of the basis of reality?

This hope surely flowed from the history that I've just mentioned – the long-standing clash between two metaphysics, two imaginary worlds – and the rivalry that inevitably developed between their sponsors. Which method – which profession – which vision of life – ought we to back? Essentially, this was seen as an issue about what we can put our trust in, what truth we think is the deepest. Traditional trust in the theological account was gradually weakened by wars of religion, which gave it bad political associations and also raised questions about what it really meant. Meanwhile, Newton developed the physical story in a clear, simple form, making it look quite plausible that, as Alexander Pope put it, 'God said, Let Newton be! and all was Light.'[1] For a time, both stories could be accepted together.

But, as the physical story grew increasingly dominant in ordinary thought – as thinkers drifted towards the idea that Matter was the only reality – some people aspired to tidy up metaphysics by getting rid of the mental story altogether. Materialists, such as Marx, evolved quite subtle devices for doing this. But it was hard to do it without losing the mind's more obvious products, such as science – and Marxism – itself. Lately, however, mind-haters have grown too impatient to bother with these subtleties, so they now attack the problem, more alarmingly, with a meat-axe, declaring that mind itself simply does not exist. All that really exists is (it seems) physical matter. Thus Francis Crick (Mr DNA in person) kindly shares with us what he calls his Astonishing Hypothesis, namely that:

> 'You,' your joys and your sorrows, your memories and your ambitions, your sense of personal identity and free will, are in fact no more than the behavior of a vast assembly of nerve cells and their associated molecules … [Then, referring to *Alice in Wonderland*, he sniffs] 'You're nothing but a pack of neurons' … [And finally, reverting to academic language …] to state the idea in stronger terms … [t]he scientific belief is that our minds – the behavior of our brains – can be explained by the interactions of nerve cells (and other cells) and the molecules associated with them.[2]

This is, incidentally, a fine example of the special reductive use of the word 'explain' in materialist rhetoric, where 'to explain' commonly means 'to translate into the terms of my own metaphysical theory'. Daniel Dennett's vast book *Consciousness Explained* is perhaps the largest, and surely the most intricate, example of this technique in recent times. But the device is common, and is often used, as it is here, to convince us that our own sense of self is simply an illusion: nothing is real except the brain cells.

This unintelligible view is, however, now put forward quite widely as being the official message of Science. I shan't say much about that campaign now because I have lately written a whole book (*Are You an Illusion?*) enquiring what on earth this suggestion could possibly mean, and asking in particular just *who* it is that is supposed to be being thus deluded. But perhaps we should now ask an even sharper question. If we have no minds, where did we get all this scientific knowledge which we apparently have? How do we know so much about things like brain cells if our minds are, and have always been, mere empty jugs, conveying pure illusion?

For instance, Crick claims to have got rid of our memories and our ambitions. But how could people who never remembered anything or formed any projects think scientifically? Science is, after all, a product of the human mind. It is simply one way of reasoning among a welter of other ways that we all use. It is not a magic cure-all that can bypass our other capacities. It is just one set of tools, quite recently added to humanity's enormous tool-kit, to do a particular range of jobs. Why should anybody think that it would give us a short-cut to the whole of reality?

Machine-magic

This idea does indeed seem so implausible that you may well begin to think I am making it up. Can it actually be true that reputable scientists are taking such an indefensible line? Well it can, and this happens because of something which I mentioned much earlier: namely, the immense power of the human imagination.

Symbols – images – visions – dreams – dramas: these are indeed the scene of all our deepest thinking. They are where real understanding goes on, and the sketchy summaries of it that we eventually formulate as academic subjects, such as the sciences, are only their shadows. When these ideas are widely endorsed we call them intuitions and treat them with deep respect. But the trouble is that, where these visions represent an idea in violent colours, our reasoning sometimes finds it impossible to resist them. And during that crucial epoch when modern science was dawning, the dominant symbol shaping all speculation about it was the vision of something that was then quite new – *machinery*, which meant, essentially, clockwork. Clocks, watches and automata were the miracles of the age.

In a startling manner, the universe began to appear as a vast clock, and everything within it as a cog or a wheel in that clock's mechanism. It's no wonder that learned people were dazzled by that tremendous image. They were deeply and understandably impressed – perhaps traumatized – by the sudden appearance of automata, the skilfully devised clockwork figures of people and animals which acted as if they were living beings. That experience was surely what made Descartes and his followers find it plausible to say that actual animals were really automata too – physical items wound up to act in a life-like manner without any real feeling. (This vision of them still persists; there are plenty of learned people today who think it is rational to assume that non-human animals are not conscious.) And as time went on, this imagery was gradually spread to form an explanation for the whole of life – an explanation that would make mind and consciousness completely unnecessary.

The interesting thing here is that nobody seems to have seen that this mechanistic vision couldn't work. Real, literal machines don't appear in the world on their own like this. They don't grow on trees. Machines can only be produced by living minds. Some conscious being always has to invent them. In fact, they are devices framed to fulfil human intentions, and anyone who wants to understand a machine has to find out what the intention behind it is. So machine-imagery may often be useful, but it is not a way of getting rid of teleology – explanation by purpose – from our understanding of the natural world. In fact, it only makes the centrality of purpose more explicit.

Teleophobic mechanism

Modernizers, however, did want to get rid of teleology because they thought it involved explanation by the will of God. Here they were mistaken. The word only means explanation by *function*, by the purpose that things serve, as we explain the heart or the liver by finding what they do for the body. Aristotle, who launched this form of explanation, was not thinking at all in terms of a creative God; indeed the Greeks had no notion of such a being. He had simply noticed these distinctive functions, and marked them by calling such effective body parts *organa*, or tools – tools of nature, not tools of God. This idea then proved so useful that it became central for the dawning science of biology, and living things are still described as 'organisms'. The idea is, of course, also relevant when the item that we want to explain is actually a machine.

It is odd to see how confidently people have ignored this connection, how readily they still use machine-talk to get rid of explanation by purpose. An example of this casualness which particularly surprised me comes from John Searle, usually a very shrewd and careful discriminator between metaphor and literal language. Asking whether artificial intelligence is possible, Searle sets up the following brief dialogue:

'Could a machine think?'

The answer is, obviously, yes. *We are precisely such machines.*

'Yes, but could a man-made machine think?'[3] (Emphasis mine)

and he goes on to argue the point, as if some machines were man-made and some were not. But all this is surely only metaphor, leading to empty conclusions. People are not machines.

What seems to be happening here is that machine-imagery has somehow worked its way into our thinking on these subjects by its own dramatic force, very much as battle-imagery worked its way into Christian thinking, and the imagery of selfishness has worked its way into modern talk about genetics. Machine-imagery has become so habitual – indeed so 'automatic' – that people scarcely notice it, let alone criticize it or see it as dangerous. But it is fast becoming the fall-back option in our culture, and we need to see how this popularity can have arisen.

The new divinity

This popularity became possible largely because of a chance of timing. Just at the epoch when educated people in the West were beginning to lose confidence in religious answers to big questions, and in the whole idea of an ordered world, this new oracle, the MACHINE – not yet discredited by failure – appeared on the horizon, claiming final authority to answer them instead. Ideas of the mysterious, perhaps infinite powers formerly attributed to God, now gradually moved over and were credited to machines – not, of course, to existing machines, but to their heirs, their promised future successors. And meanwhile the rest of us, when we fall into any difficulty, have been told to seek help and counsel, not from God but from the current representatives of those physical powers – from scientists and technologists.

This raises interesting questions about the kind of help these representatives can offer. No doubt the earlier religious beliefs had faults, but their world-picture did at least represent ideals close to the hearts of the people who trusted in them – ideals such as love, charity, purity, reverence, courage, loyalty and compassion. These ideals were complicated and sometimes conflicting, but they were real, visibly important elements in ordinary life. They placed us in a whole community of others who shared them. By contrast, the ideals behind the image of super-mechanized life which have succeeded them are not complicated at all; they impress us by a single dazzling quality – effectiveness. To make up for this thinness they offer us the promise of infinite – but also indefinite – wealth and success some time in the future.

This offer has naturally proved attractive. Today, when we come across something puzzling in the natural world, our academic super-ego tells us to look only for the mechanism that makes it happen. We are sure that this will be the real explanation behind the surface phenomena, so it will give us the path that we should follow next. And the key to it will always be found through Science, which is in touch with solid reality – matter – the substratum behind surface appearances.

Immanence and transcendence

But is this the only kind of explanation that we need? Perhaps we should ask here just what the idea of divine creation has meant for us so far. That idea always had two aspects – a transcendent side, a powerful creator working outside Nature, and an immanent side, natural creative tendencies within the created beings themselves.

According to current ideas, the Darwinian revolution has removed the transcendent creator without any difficulty, despite that creator's importance in our social and personal life. But the neo-Darwinians saw that the *immanent* creator – growth itself, which is a central force in all earthly organisms – was real and must be treated with respect. We must still ask, why is it that an acorn goes through the whole troublesome process of becoming an oak tree, rather than just turning pink or rotting away at once?

Notoriously, Darwin himself suggested just one simple reason for this surge of directed energy – natural selection between competing alternatives among its ancestors, ensuring that the fittest always survived. Thus acorns that became oaks left more descendants than ones that didn't. And of course, given the present set-up – given a species already in business – that is perfectly true. But could it account for all the steps between primitive organisms and a prosperous species of trees? Does it explain why development has followed one pathway rather than another? Does it (for instance) explain why our own species has developed the capacity to do advanced mathematics, something which surely cannot at first have been much use in providing extra descendants?

Darwin thought not. He said plainly that natural selection could not possibly be the whole explanation for the actual vast, inexhaustible expansion of living organisms, and he repeated that point quite crossly in edition after edition of the *Origin*. 'Great' (he wrote) 'is the force of misrepresentation.'

On this matter, however, most of his followers have just ignored him, and still do. Fascinated by the simplicity of his first idea, they assume that natural selection is indeed the sole explanation for evolution. And when they can't find any selective advantage to be drawn from a given change, they simply make one up. But this reliance on a single explanation surely cannot be right because it works on far too small a scale. No doubt, given present conditions, a single fitness competition of this kind could work. But that is only possible because now the two competing tendencies, and the whole background of conditions that they have to adapt to, are already provided. There is no need to decide which direction to go in next.

Without any transcendent Creator or programmer, it is not obvious how this guiding background could be generated. If we want to use this slow and chancy process of adaptation to account for the whole vast ordered growth of earthly organisms, we shall surely need, not only a far longer time than is actually available, but also some explanation of how the particular paths that have been taken have been chosen. It is not that we necessarily require a personal director. But what evolution does need is a signpost – an indicator of direction.

Convergence, clan formation and chlorophyll

Plainly, evolution has not just diverged constantly into every possible direction, following the chance distribution of alternatives at each particular juncture. That would simply have led to confusion. Instead of this, an immense procession of species have managed to co-operate in their own development, sorting themselves out in orderly fashion into a limited number of kinds, kinds in which various particular evolutionary 'inventions' – chlorophyll, bony skeletons, spider-silk, shells, fur and the like – each dominate their own province and have become stabilized.

Other alternative possible forms such as wheels (which we might think more promising) have occurred occasionally but they have never caught on. And it is clear that this sorting-out has not been done simply by heredity and natural selection, because there are a crowd of cases of convergence – cases where organisms from different sources independently home in on the same solution for the problem that they share.

The best-known case of this convergence is the camera-eye, an ingenious but very complicated device which we ourselves are lucky enough to possess, and which at least eight other unrelated creatures share with us, including ones as distant from us genetically as the common octopus. All over this immense realm of possibilities, this general pattern of convergence is widespread. In fact, it is normal.

What has emerged, then, is that certain forms, certain possibilities, are always in some mysterious way more attractive to developing organisms than others. Simon Conway Morris compares this discriminating pattern of development to a map which makes it possible to find a single island – say, Easter Island – in a vast unmarked area of ocean:

> Convergence occurs because of 'islands' of stability, analogous to 'attractors' in chaos theory … Convergence gives us a clue as to how the metaphorical Easter Island is located, but just as this Pacific Island is surrounded by oceanic wastes, so too

perhaps are the 'islands' of biological habitability. I suspect that not only will the bulk of biological space never be occupied, but it never can be either.[4]

This unevenness – this greater drawing-power of some alternatives over others – suggests a much less simple pattern of development than the sweeping, one-sided one that neo-Darwinists have so far been demanding. It is surely more realistic. To quote Conway Morris again:

it is now widely thought that the history of life is little more than a contingent muddle punctuated by disastrous mass extinctions ... Yet, what we know of evolution suggests the exact reverse: convergence is ubiquitous and the constraints of life make the emergence of the various biological properties very probable, if not inevitable.[5]

In short, instead of a random sequence of separate competitions, each settled by local chance factors, we now see a slow but steady drift of waves in a certain definite direction – a drift evidently responding to particular pre-existing tendencies, a kind of constant pattern of trade winds that brings with it its own sort of order. Evidently, the whole thing is going somewhere.

In this context, teleological questions begin to make sense again. We no longer have to insist that evolution must be a random and pointless process. That melodramatic assumption

– which has, for no particular reason, been treated as being particularly sage and economical – is really no more than a theorizer's arbitrary preference like any other. We can now reasonably ask, what is the reason for this particular creature's tendency to burrow, or to fly, or to build shells, or to form universities? We can now wonder, which way of living is *life itself* aiming at here?

Evolutionists do, of course, already ask many questions of this kind, often without bothering to feel guilty about their teleological bias. And among those many questions, the one already mentioned about the camera-eye is typically significant. Why (we ask) do so many different creatures invest their scarce resources in developing this distinctly expensive kind of eye? And the answer, as Conway Morris points out, is that this is only part of a much wider tendency among animals to concentrate on developing their perception, their consciousness, observation and understanding *as ends*, even at the expense of their purely physical capacities.

In fact, it turns out to be an objective fact about life that increased awareness of the world is an important end in itself for living creatures. That is surely why there is an immanent tendency in all life to increase awareness, whether a Supreme Being has been present to put that tendency there or not. The sense that such ends have real value is not a mere chance consequence of local religious – or irreligious – opinions. It is one of the deep roots out of which the great trees of the various religions have grown in the first place. And it is at the root of science as well.

Scientism itself has of course sprung from these same roots, this same knowledge-hunger, this same urge to get a deeper understanding of the world around us. Like other intellectual enterprises, scientism always aspires to this deeper understanding. But it does so in unusually exclusive terms. It regards the physical sciences not just as important aspects of knowledge but as its centre, filling the whole intellectual heaven. From the scientistic point of view, other forms of scholarship – history, poetry, philosophy, ethics, theology, the arts, the social sciences, which so far have all seemed to be necessary aspects of life – now appear as subsidiary approaches, provisional suggestions, minor arts that need confirmation from physical science – indeed, if possible, from physics itself – before their results can be treated as conclusive.

Knowledge and value

Instead of these familiar ways of thinking, what is expected to save humanity today is a special kind of intellectual advance, a particular increase in skills, perhaps essentially electronic skills, which is expected to bring us together with the machines and (as we shall shortly see) to lead us to salvation at the Singularity.

This confident reliance on future machinery – this mechanist creed, so powerful today – cannot exactly be described as a 'religion of science' because it lacks the personal passion that has always been central to religious salvation. It offers us nothing like the familiar individual love and gratitude that souls have felt towards Jesus or Mary or Pan or Apollo. Nor, indeed, does it give us moral guidance of the kind that we expect from religion. It does not tell us what sort of life we should live when we are outside the laboratory. Yet it evidently claims a powerful moral authority on matters that it considers do come within its precinct. And its main command is simply that we should go on increasing our scientific knowledge.

This is not new. Whenever the demands of scientific research conflict with those of ethics, high-minded theorists can always be found to explain why research should always come first. Thus that impressive prophet Jacques Monod, explaining the relation between science and morality, tells us:

> In order to establish the *norm* for knowledge the objectivity principle defines a *value*; that value is objective knowledge itself … The ethic of knowledge that created the modern world is the only ethics compatible with it, the only one capable, once understood and accepted, of guiding its evolution … As for the highest human qualities, courage, altruism, generosity, creative ambition, the ethic of knowledge both recognises

their sociobiological origin and affirms their transcendent value in pursuit of the ideal it defines ...

Finally, the ethic of knowledge is, in my view, the only attitude which is both rational and resolutely idealistic, and on which a real socialism might be built.[6]

In short, the only thing that has any real value is held to be the pursuit of truth – not just any old truth but truths about the physical world, found out through physical science. Other human ideals and standards are not considered to be valuable except in so far as they serve this purpose.

Claims like this have, of course, been made before. Both Plato and Aristotle wrote that certain particular forms of spiritual activity are more important than all other human concerns. But then, they supported those claims by wide and impressive accounts of our inner needs and destiny. Today, nobody has made any such deeper diagnosis of the human situation. It is simply being taken for granted that what we centrally need is just more physical knowledge and more machines built on that knowledge. This vision of 'the future' is held to be all-sufficient, and its claims are supported only by increasingly hopeful descriptions of this prospect.

16

Missing persons

All this powerful rhetoric illustrates the way in which familiar visions – well-known and potent images – can shift our morality and take the place of the plain facts before us. That is what I meant when I said earlier that the human imagination has far more power than we realize. Of course that imagination also has great power for good. It gives us our new panoramas, new insights which are the source of all our fresh ideas and inventions. But if we don't want these new developments – if we are anxious to *resist* change – imagination also provides us with an escape-hatch by making a noise somewhere else. In strengthening our resistance, it plainly often has power to set aside all regular processes of argument.

That is why people who are in the grip of a favoured vision often become quite immune to regular argument, and it is why their favoured vision is often rather a primitive, childish one. Thus, battle-imagery has an obvious primitive appeal, and the childish emotive image of 'the nanny state' is enough to stop many people attending to the alarming facts of climate change altogether. Similarly, seeing the living world as simply a mass of machinery and ourselves either as engineers or as cogs within it obscures the centrality of our own independence and spontaneity. Mill, arguing for free discussion, stresses that centrality:

> really is of importance, not only what men do, but what manner of men they are that do it. Among the works of man, which human life is rightly employed in perfecting and beautifying, the first in importance surely is man himself. Supposing it were possible to get houses built, corn grown, battles fought, causes tried, and even churches erected and prayers said, by machinery – *by automatons in human form* – it would be a considerable loss to exchange for these automatons even the men and women who at present inhabit the more civilised parts of the world, and who assuredly are but starved specimens of what nature can and will produce … *Human nature is not a machine to be built after a model and set to do the work prescribed for it, but a tree*, which requires to grow and develop itself on all sides, according to the inward forces which make it a living thing.[1] (Emphasis mine)

Is it not extraordinary that, a mere couple of centuries after Mill wrote this, sane and well-educated people should reject his inspiring vision of the tree altogether and should insist instead that they themselves will soon become mechanized? Is it not odd that they should be taking positive pride in claiming that they are already becoming ever more machine-like, and should predict that those who will rule the earth after them will be essentially machines? We shall have to ask more questions about the meaning of this doctrine shortly. But first, according to my usual method, I want to look back and ask where it came from.

The dawn of mechanolatry

This dream of mechanizing people by way of anthropizing machines seems to have originated in the early nineteenth century, in response to the appearance of increasingly sophisticated calculating devices. Since these machines produced intellectual results rather than just physical ones, they could be seen as independent thinkers. But it is still remarkable how quickly, and how readily, this dramatic conception of their status caught on and grew into an influential myth. For instance, in 1847 the editor of *The Primitive Expounder*, R. Thornton, responded thus to the recent invention of a four-function calculator:

such machines, by which the scholar may, by turning a crank, grind out the solution of a problem without the fatigue of mental application, would, by its introduction into schools, do incalculable injury. But who knows that such machines when brought to greater perfection, may not *think* of a plan to remedy their own defects and then grind out ideas beyond the ken of mortal mind?[2]

Here we see already two of the many disconnected responses which constantly greeted these inventions – first, a fear that they will undermine existing skills, and then a hope that they will bring unlimited new capacities to replace them. This second proposal marks the onset of the new potent myth. Here the author is already seeing these objects, not as new tools for human use, but as new allies or advisers, new sources of power, original thinkers who can outdo and save the people who invent them. Indeed, he is already treating these creatures as superhuman, since he assumes that they can 'remedy their own defects', something which human beings notoriously can't do. It is surely a remarkable spasm of confidence to credit our unknown successors with this crucial power of self-correction.

And so it was that, just halfway through the nineteenth century, at the very point when educated Britons were congratulating themselves on becoming free from superstition because they were beginning to lose their confidence in God, a fresh set of supernatural beings, conscious machines, who would be able to replace that God and Nature itself as the new substitutes for teleology, began to appear. Some sense of this is expressed in Samuel Butler's novel *Erewhon*, where Butler, arguing that predictions about conscious machines are plausible, writes:

> There is no security ... against the ultimate development of mechanical consciousness, in the fact of machines possessing little consciousness now. A mollusc has not much consciousness. Reflect upon the extraordinary advances which machines have made during the last few hundred years, and note how slowly the animal and vegetable kingdoms are advancing. The more highly organised machines are creatures not so much of yesterday but of the last five minutes, so to speak, in comparison with past time.[3]

Why did Butler apparently take it for granted, not just that machines are advanced living creatures, but that they form one of the species supposedly competing in the current Darwinian – or rather Spencerian – race for the survival of the fittest? Is this not somewhat supersititious? And why did he write – without ever mentioning the changes in human behaviour that actually produce these machines – as though this new mechanical tribe was driven, like its competitors, wholly by its own inner evolutionary force? Was his neglect of their social causes in human life – causes that had produced the whole Industrial Revolution – intended to be swallowed whole as literal truth?

Machines taking control

Butler may well not have intended it so. *Erewhon* is in general so far from literal, so subtly ironic and satirical, shooting arrows from so many different directions at the Darwinian vision, that we have to interpret it for ourselves. Many readers, however, have indeed swallowed this idea whole. Among them was no less a figure than Alan Turing, who wrote in 1951 (in a paper called 'Intelligent machinery, a heretical theory'), 'Once the machine thinking method has started, it would not take long to outstrip our feeble powers … At some stage therefore we should have to expect the machines to take control, in the way that is mentioned in Samuel Butler's *Erewhon*.'[4]

This is the myth. But how are these beings expected to set about taking control? How are they expected to make the leap from success in a few, carefully selected theoretical puzzles, specially adapted to their peculiar powers, to a general management of human life? How (in fact) are they to move from the role of servants to that of employers, from calculating about chosen, highly abstract problems to a usable, practical grasp of the whole human scene?

We know that, as things go now, people who have to deal directly with large practical questions, such as how to stop a war, or to control a drought, or to bring two quarrelling factions into agreement, do not do it simply by being very clever. We know that they need to have a great deal in their minds besides a high IQ.

Usually indeed, the first thing that these administrators need is not so much stronger gifts as a lifetime's experience of existing facts – facts which, for all anyone knows, may be completely miscellaneous. And, for a start, they need something which no machine could ever get, namely that actual experience. They must have the memory of a lifetime spent among detailed human transactions, a memory angled so as to bring out the special difficulties of this new task and provide attitudes that can resolve them. And, since no-one knows in advance what those attitudes will be, there is no way in which machines could be programmed to produce all this. In fact, if it were possible for someone to tell a machine that this was what was expected of it, and if that machine were able to give a rational reply, its answer would have to be 'NO!'

Could the machines themselves somehow get us out of this difficulty? Of course people have tried to use them in that way. Those who now work on these problems use machines to make sure that they get the best possible information for locating the crucial issues and finding good solutions.

But the trouble comes long before this, at the point where they must choose the general direction of their policy – where they must decide what they are aiming at and what questions to ask. They cannot hand this work over to yet more machines. They must still be guided by their own individual skills and characters, their private beliefs, their personal attitudes, their feelings, their education, their peculiar experiences, their own sense of values. The practical dilemmas confronting these people are not standard fixed puzzles like chess problems. They are wide seas of problems, indefinite clusters of conflicting practical considerations which, as far as anybody knows, might lead anywhere. Thus these much-stressed administrators work in a kind of general bewilderment in which no machine could ever possibly find itself.

17

Oracles

The search for certainty

When human beings get lost in this way, their commonest response is to look for an oracle – an agency that will take the next decision for them. In past times, oracles were most often expected from the gods. Thus Croesus, king of Lydia, asked the Delphic oracle whether he should attack the Persians, and was told that, if he did, he would destroy a great empire. (This was, of course, correct, only the empire was his own.) Somewhat later, people who had lost confidence in these theistic methods took to looking for their oracles mainly in books, such as *Leviathan* and *Das Kapital*. But only quite lately have they thought of finding them instead in machines – that is, by looking them up in Google.

This option has obvious attractions. For a certain range of questions, machines do indeed give reliable answers, and many of these are questions which humans themselves cannot answer at all. But these machine-digestible questions are all formal ones of a certain kind, mainly ones that can easily be reduced to terms of quantity. And these questions don't figure largely among the urgent practical problems which human administrators have to deal with.

Instead, there is usually a vast background of less malleable, more miscellaneous, non-quantifiable but still crucial problems, a background that includes not just moral dilemmas such as 'Do I have to keep these rules?' or 'Is torture always wrong?' but also many miscellaneous factual ones, such as 'How can we make a torture-based prison system work in spite of objections?' In recent decades the hope, the vision, the dream of extending machine competence to cover this whole worrying mass of background queries has become so strong that it has led people to form what, in times past, would surely have been called a bizarre superstition, namely the belief that, at some future date, the advantages of non-human intelligence will somehow be combined with those of ordinary human thought, so as to provide reliable oracles for all these difficulties. And the future point at which this is expected to happen is now known as the Singularity.

PART FOUR

SINGULARITIES AND THE COSMOS

18

What kind of singularity?

Alarming futures

So what is this singularity, if not just an extension of Google?

In ordinary speech, the word 'singularity' only appears in the harmless, everyday sense of strangeness, oddity or uniqueness. But, outside that context, it has gathered a number of specialized more or less mathematical meanings such as 'a point where all parallel lines meet', and 'a point where a measured variable reaches unmeasurable or infinite value'. Recently these concepts have been collected and moulded into the idea of the 'technological singularity', which is the name now given to the phase of profound, future change at which machines are expected to become more intelligent than human beings, machines having become cleverer and humans more mechanized. Possibly, too, people and machines may then simply be combined.

This idea is now being eagerly supported. In California there is a thriving Machine Intelligence Research Institute (previously called the Singularity Institute) which is occupied, among other things, with AI safety research – that is, with finding ways to protect us against attacks by newly empowered and humanized machines that may follow from this startling event. And Ray Kurzweil, self-denominated futurist and founder of the Singularity University, expects that event to happen quite soon, perhaps about 2045.

By contrast, Noam Chomsky, the philosopher who has been most carefully thinking through this whole concept of artificial intelligence ever since that phrase was invented in 1956, describes the singularity idea roundly as fiction. Being asked whether we ought to be alarmed that the appearance of machines cleverer than humans might destroy our species, he replied soberly:

I think we should be concerned about the end of our species, but not for that reason. We should be concerned about it because we are very busy destroying the possibility for peace and survival; we should worry about that, like the latest IPCC report; but the singularity stories are science-fiction.

When we talk about machines, it's kind of misleading; people have a feeling, it's this cute little robot like we had in R2-D2.

The reference to machines is a reference to programs, the program that you put into the device that sits on your desk. The device itself is of no use for anything other than maybe holding down papers. It's a paperweight ... but it can execute programs. A program is a theory ... [and about that theory] you ask the same questions that you do about any other theory; does it give insight and understanding? (Chomsky 2013, 4)

This realistic line of talk blocks at once the flood of life-conferring imagery, the incipient personification which has led so many people, from Samuel Butler onwards, to inflate these imagined future machines into outsize quasi-humans, whether bad or good. Chomsky thinks this whole story is just a misleading fancy that distracts us from our real troubles, whether it paints the machines as angels coming to save us or as demons out-competing us into extinction. Without that bogus drama (he says) the alarming – or tempting – idea of an endless rise in non-human cleverness would never have got off the ground.

And indeed that idea is not a clear one. As Alan Turing, who is often widely regarded as the patron saint of all these enquiries, wrote in his famous paper on whether machines can think, the question of whether machines can think is 'too meaningless to deserve discussion'.[1] Similarly, Chomsky remarks at the start of the same discussion that, without that dramatic imagery, the whole concept of artificial intelligence is hopelessly obscure:

> What the term is usually used for is the effort to program machines ... to mimic certain aspects of human behaviour. That can be understood as a scientific project, like we might try to construct a model of, say, insect navigation, *to try to understand insect navigation better*, or it could be an effort to produce something of utility, like a robot that will clean your house better. Those are basically the two thrusts of artificial intelligence. The work in these fields over about 60 years *has not given any insight to speak of into the nature of thought, the organization of action*, and I don't think that's very surprising. (Emphasis mine) (2013, 1)

It is indeed not surprising because these two aims don't converge at all; they are actually in conflict. Trying to produce useful gadgets involves imitating certain aspects of current human behaviour, with adjustments that are meant to make those gadgets more useful, not to deepen our understanding of the behaviour. And since these imitative inventions rest on our present vague ideas about how thinking works – something about which at present we know very little – this 'imitation game' cannot possibly teach us anything more about thought.

Thus, although work on developing things like self-driving cars may well be practically useful, it contributes nothing to psychological research. It does not help us at all to understand self-driving people, which is our real problem. So it cannot enable us to make sense of the wild project of joining humanized machines with mechanized humans to generate useful hybrids.

Singularities and memes

Is it perhaps then a waste of time to discuss such wild projects here at all? Scholarly standards would once have made it seem so. It would once have been thought necessary to keep such things in cold storage until they had been properly tested. But today the internet floods us with so much varying chatter about all sorts of new suggestions that I think this segregation has become unworkable. The Singularity is at present being widely enough discussed – and the fact of this discussion seems itself significant enough – to make it sensible to say something about it here.

Indeed, on this point talk about the Singularity seems to me to stand on much the same footing as talk about memes. Some thirty years ago, memes were for a time regarded seriously as scientifically respectable entities, essentially as genes of culture, an important element in the idea of social evolution. Indeed, for a time memetics was seen as a useful branch of evolutionary biology. But, as people looked into the matter, it soon became clear that the analogy between genes and memes had never been thought through properly at all.

Culture is, in fact, a radically different kind of thing from the organisms that employ genes. Cultures, and their elements, are shifting shapes of tradition, not fixed entities that reproduce in any standard way. So they have no lasting standard parts that could possibly correspond to genes. Moreover, it soon became obvious that the cultural items that were first named as examples of memes – popular songs, stiletto heels and the Jewish religious laws[2] – were a completely mixed bunch, none of which had anything like the kind of fixed, permanent form that is supposed to characterize genes. Besides this, since that time a steadily increasing grasp of epigenetic influences has made it clear that genes themselves are nothing like as central in the evolutionary story as they were believed to be when the idea of memes was first launched.

Evolutionists therefore soon lost interest in this kind of argument and the idea of memes died out quite quickly from serious biology. The word 'meme' survives, being used occasionally in sociology, but its sense is vague. It seems to me (though I am no qualified futurist) that the language now associated with the Singularity surely faces a similar fate. If, as may reasonably be expected, nothing particular happens in 2045 or soon after that would give a sense to it, this whole prophecy will probably soon be forgotten, being replaced by whatever suggestion next takes the public fancy.

The end of comprehension?

To return then to the idea of Singularity itself – what does this prophecy mean? Different prophets vary in their views about the form that this startling change is to take, but there is general agreement that it will arise from a convergence between 'artificial intelligence, genetics, robotics, virtual reality, biotech and nanotechnology'.[3] It may be concentrated in an extremely small device, or set of devices, so that 'what currently fits in your pocket will eventually fit into a single blood cell and connect directly with your body'.[4] But one central theme pervading these suggestions is that it is something completely new – existing human minds will not be able to understand it. Thus, in 1958 Stanislaw Ulam wrote about a conversation with John von Neumann in which von Neumann said that (in Ulam's words), 'the ever-accelerating pace of technology ... gives the impression of approaching some essential singularity in the history of the race beyond which *human affairs, as we know them, could not continue*' (emphasis mine).[5] Von Neumann's alleged definition of the singularity was that it was the point beyond which 'technological progress will become *incomprehensibly* rapid and complicated' (emphasis mine).

The new dawn

This is strong language. Von Neumann's pronouncement can, of course, be taken simply as a dirge, a final requiem mass for all human attempts to understand technology. Many theorists, however, reject this conclusion of the matter. Their faith is that the final failure of existing human intelligence will be more than balanced by the co-operation and ultimate dazzling success of our non-human heirs, the machines. Thus I.J. Good viewed this great event as heralding a new dawn, a positive feedback cycle within which minds will make devices to improve on their predecessors, devices which, once started, will rapidly surge upward and create super-intelligence:

Let an ultraintelligent machine be defined as a machine that can far surpass all the intellectual activities of any man, however clever. Since the design of machines is one of these intellectual activities, an ultraintelligent machine could design even better machines; there would then unquestionably be an 'intelligence explosion' and the intelligence of man would be left far behind. Thus the first ultraintelligent machine is the last invention that man need ever make.[6]

And, as we have seen, Kurzweil predicts that this change will indeed occur soon, probably at some point around 2045, after which:

artificial intelligence will become sufficiently powerful and generalised that it can start to improve its own design in a recursive fashion. This will result in a runaway effect: from that moment the intelligence of AI will expand with ever-accelerating rapidity, with consequences beyond our ability to predict. It will swiftly reach a point, says Kurzweil, '[where] the pace of change is going to be so astonishingly quick that you won't be able to follow it unless you enhance your own intelligence' through technology.[7]

19

Can intelligence be measured?

Who understands what?

Readers will notice how these predictions that the future will soon become quite unpredictable are followed at once by confident prophecies about specific results that will come of the change. That, however, is by no means the most disastrous fault in this way of thinking. The whole original conception of 'intelligence' itself is hopelessly unreal.

Prophecies like these treat intelligence as a quantifiable stuff, a standard, unvarying, substance like granulated sugar, a substance found equally in every kind of cake – a substance which, when poured on in larger quantities, always produces a standard improvement in performance. This mythical way of talking has nothing to do with the way in which cleverness – and thought generally – actually develops among human beings. This imagery is, in fact, about as reasonable as expecting children to grow up into steamrollers on the ground that they are already getting larger and can easily be trained to stamp down gravel on roads. In both cases, there simply is not the kind of continuity that would make any such progress conceivable.

The trouble here is that human life actually contains many different sorts of cleverness or intelligence because we are called on to attempt many different sorts of understanding, and every particular challenge needs the particular sort that is appropriate to meet it. This is the point that Noam Chomsky made in insisting that what we call 'intelligence' is really many intelligences, a collection of capacities, 'cognitive modules' suited to deal with particular kinds of subject-matter. Thus our language-learning capacities form a distinct nervous system, one that is parallel to our visual system or our immune system. The linguistic one is amazingly flexible in dealing with human languages and cyphers, but it is still specialized. It could not work if it was confronted with an alien communication constructed on totally different principles. As Colin McGinn says, in that case 'The task and the tool are not made for one another … It's like trying to crack nuts with a feather duster.'[1]

In fact, understanding something is not actually a simple, brute-force operation, like digging a hole in hard ground. It does not just need stronger arms or a sharper spade. It is usually much more like trying to find the main jigsaw puzzle into which the isolated piece that you are now holding fits. To do this, you have to look for a context where that piece makes sense.

For instance, you hear someone repeatedly saying the word 'aqua', but you don't know what language they are speaking. Only when it strikes you that they may be speaking Italian do you properly hear the word as 'water', and with it grasp the whole background that explains their meaning. To do this you must either know Italian already or have a good enough grasp of human behaviour to be able to interpret gestures, to see the general pattern of what they are trying to convey. And even when we do have this previous prompting, we still need to go through the troublesome activity of interpreting; we still need to put the clues together in a way that sketches out that dominant pattern, which is often much harder than it is in this example.

What makes this possible is not just abstract sharpness, a general power that might be measured and recorded in terms of IQ. It is much more the direction of our own interests, our keenness to perceive particular aspects of life and to pick out what is important there. In fact, what makes us call a person specially intelligent is far more often this perceptiveness, this power of distinguishing what really matters in a certain field, than a mere general capacity. And people's different interests make them differ on this point much more than might have been expected. It is far more a matter of motivation than of capacity, and of course machines – poor things – don't have any motives at all.

We recognize this divergence of interests all the time when we are trying to find suitable people for different situations. Thus Bob may be an excellent mathematician but is still a hopeless sailor, while Tim, that impressive navigator, cannot deal with advanced mathematics at all. Which of them then should be considered the more intelligent? In real life, we don't make the mistake of trying to add these people's gifts up quantitatively to make a single composite genius and then hope to find him. We know that planners wanting to find a leader for their exploring expedition must either choose between these candidates or send both of them. Their peculiar capacities grow out of their special interest in these topics, which is not a measurable talent but an integral part of their own character.

In fact, the word 'intelligence' does not name a single measurable property, like 'temperature' or 'weight'. It is a general term like 'usefulness' or 'rarity'. And general terms always need a context to give them any detailed application. It makes no more sense to ask whether Newton was more intelligent than Shakespeare than it does to ask if a hammer is more useful than a knife. There can't be such a thing as an all-purpose intelligence, any more than an all-purpose tool. And the kinds of cleverness that machines display in calculations are, as we have mentioned, all formal, specialized and essentially quantitative. They are skills belonging strictly to the left hemisphere of the brain – the part that deals with specialized, technical problems – so they are not continuous with the ones that have their cerebral basis in the right hemisphere, in thoughts about the wider context and the rest of our lives. Thus the idea of a single scale of cleverness, rising from the normal to beyond the highest known IQ, is simply a misleading myth.

It is unfortunate that we have got so used today to talk of IQs, which suggests that this sort of abstract cleverness does exist. This has happened because we have got used to 'intelligence tests' themselves, devices which sort people out into convenient categories for simple purposes, such as admission to schools and hospitals, in a way that seems to quantify their ability. This leads people to think that there is indeed a single quantifiable stuff called intelligence. But, for as long as these tests have been used, it has been clear that this language is too crude even for those simple cases. No sensible person would normally think of relying on it beyond these contexts. Far less can it be extended as a kind of brain-thermometer to use for measuring more complex kinds of ability. The idea of simply increasing intelligence in the abstract – rather than beginning to understand some particular kind of thing better – simply does not make sense.

This means that prophecies like these two made by Vernor Vinge about the post-singularity world are not descriptions of any possible kind of future event. Vinge writes:

We will soon create intelligences greater than our own. When this happens, human history will have reached a kind of singularity, an intellectual transition as impenetrable as the knotted space-time at the center of a black hole, and the world will pass far beyond our understanding ...

Within thirty years, we will have the technological means to create superhuman intelligence. Shortly after, the human era will be ended ... I think it's fair to call this event a singularity. It is a point where our models must be discarded and a new reality rules. As we move closer and closer to this point, it will loom vaster and vaster over human affairs till the notion becomes a commonplace. Yet when it finally happens it may be a great surprise, and a greater unknown.[2]

These are fables, written in a vague language that describes nothing. So is this from Hans Moravec:

Our artefacts are getting smarter, and a loose parallel with the evolution of animal intelligence suggests one future course for them. Computerless industrial machinery exhibits the behavioural flexibility of single-celled organisms. Today's best computer-controlled robots are like the simpler invertebrates. A thousand-fold increase in computer power in this decade should make possible machines with reptile-like sensory and motor competence. Properly configured, such robots could do in the physical world what personal computers now do in the world of data – act on our behalf as literal-minded slaves. Growing computer power over the next half-century will allow this reptile-stage will [sic] be surpassed in stages producing robots that learn like mammals, model their world like primates and eventually reason like humans. Depending on your point of view, humanity will then have produced a worthy successor, or transcended inherited limitations and transformed itself into something quite new. No longer limited by the slow pace of human learning and even slower biological evolution, intelligent machinery will conduct its affairs on an ever faster, ever smaller scale, until coarse, physical nature has been converted into fine-grained personal thought.[3]

Future-based credulity

The most surprising thing about these manifestoes is surely their welcoming and reverent tone. If we really expected such a drastic and uncontrollable change, we would surely be frightened and would call for whatever precautions seemed possible. By contrast, these people are happy to combine their claim that we know nothing of the details of this prospect with an increasingly religious accent of respectful awe in awaiting it. Vinge's language is not at all like that of someone warning us about some ordinary danger; it is more like the voice of a pious, but unimaginative, believer foretelling the Day of Judgement. Why is he so cheerful about it? Moravec meanwhile tells his quasi-zoological story in drearily factual tones, not bothering to explain how these developments among the machines are going to affect our own species. Are we to soldier passively on, wholly directed by these machines, much as the machines are now directed by us? Or do we just go extinct?

It is surely remarkable that none of these writers strikes anything like the note of alarm which Samuel Butler struck in Chapter 24 of *Erewhon*, when he reported the Erewhonians' horror at the increasing influence of advanced machinery in their own lives – horror which, in the end, led them to get rid of the machines entirely. As Butler says, the only reason why we now regard the similar mechanization of our own culture as inevitable is because we are taking pains to make it so:

Man at present believes that his interest lies in that direction; he spends an incalculable amount of labour and time and thought in making machines breed always better and better; he has already succeeded in effecting much that at one time appeared impossible, and there seems no limit to the results of accumulated improvements if they are allowed to descend with modification from generation to generation. It must always be remembered that man's body is what it is through having been moulded into its present shape by the chances and changes of many millions of years, but that his organization never advanced with anything like the rapidity with which that of the machines is advancing. This is the most alarming feature in the case, and I must be pardoned for insisting on it so frequently.[4]

These words of Butler's are a shock because they mention what is so often overlooked – the distortion of human life that inevitably follows on people's devoting all this labour, time and thought to mechanical pursuits rather than to other human affairs – a point which Daniel Dennett stresses and which, as we've seen, also troubles Chomsky. Butler did not, like so many prophets today, simply accept this obsession with machinery respectfully, as an inevitable and awe-inspiring cosmic event. He saw it, more flatly and realistically, as resulting from deliberate, arbitrary human choices which could easily be altered. He expected it to be disastrous, again in a more realistic sense than the more depressed Singularitarians whom we have visited – not predicting that the machines will literally conquer us, but that our obsession with them must inevitably distort our lives.

This alarmed response to machine power is not just an outdated Victorian fancy. It has recently been echoed no less forcibly by Ted Kaczinski, who wrote in *Industrial Society and Its Future*:

The human race might easily permit itself to drift into a position of such dependence on the machines that it would have no practical choice but to accept all of the machines' decisions. As society and the problems that face it become more and more complex and machines become more and more intelligent, people will let machines make more and more of their decisions for them, simply because machine-made decisions will bring better results than man-made ones. Eventually a stage may be reached at which the decisions necessary to keep the system running will be so complex that human beings will be incapable of making them intelligently. At that stage the machines will be in effective control. People won't be able to turn the machines off, because they will be so dependent on them that turning them off would amount to suicide.[5]

That seems plausible enough, apart from the optimistic prediction of machines actually becoming more and more 'intelligent' – presumably in the ordinary sense of real excellence, not in the special sense that applies to efficient machines today.

So, if we believe Kaczinski and Butler, the myth that deludes Vinge is a rather ordinary form of self-deception – the delusion by which we simply attribute troubles that flow from our own folly to grand and uncontrollable cosmic forces, so as to avoid having to do anything about them.

Moravec, on the other hand, has a much more elaborate myth, one that aims to change the concept of evolution altogether. He expects machines to imitate and duplicate the course of animal evolution, arriving finally at what he clearly takes to be the inevitable aim of all such development, becoming able to 'reason like humans'. By doing this faster and becoming smaller they will then proceed on this level until 'coarse, physical nature has been converted into fine-grained personal thought'. Thus – as happened in J.D. Bernal's even bolder Utopian story, *The World, the Flesh and the Devil* – what was originally conceived as a materialist Utopia turns into an idealist one, ending with spirits that have got rid of matter altogether.

20

What is materialism?

And, by the way, what is matter?

Alarming ambiguities

That sober volume, *The Oxford English Dictionary*, which never gives variations just for the fun of it, offers us two widely different meanings for 'materialism'. On the one hand it says that this means 'the opinion that nothing exists except matter and its movements and modifications'; on the other, that it means 'devotion to material needs or desires, to the neglect of spiritual matters'.

If we wonder how the word managed to combine an abstract, rather technical, ontological meaning like this with an everyday one which – as the quotations show – carries a load of disapproval, we need only look, as we have done before in these cases, at the symbolism that our indefatigable imaginations have built around it. 'Material needs' suggests to us primarily food and sex and then the less creditable worldly interests such as riches – while 'spiritual' suggests something more uplifting, therefore good.

But it is not clear how these two moral categories can be linked with the metaphysical division of the world into matter and spirit. Indeed this connection is remarkably crude. Plenty of unquestionably bad things, such as hatred and revenge, are spiritual rather than material; plenty of material needs, such as our love for each other which leads us to touch and embrace our loved ones, are good. Thus this commonplace symbolism does not give a reliable moral compass.

That is probably one reason why philosophers have recently begun to avoid the ambiguous word 'materialism' altogether, and now speak of 'physicalism' instead. This word, as the OED kindly explains, was invented by the logical positivists of the Vienna Circle as a name for the view that 'reality is all reducible to certain kinds of physical entities' or, as Carnap put it, all science can be expressed in the language of physics.

This change plainly struck theorists as simply a neat way to wall off serious ontology from crude, everyday moralizing about human conduct. But moving from everyday speech to technical language always has its price, and in this case the price is high. If 'physical entities' means only ones that can be described in the language of physics, then everyday life simply contains hardly any physical entities at all. Physics never speaks of loaves and apples, pens and paper, men and women, bricks and mortar. It always speaks, far more abstractly, of solids and liquids, protons and electrons, vacuums and black holes.

The philosophers of the Vienna Circle were not actually trying to redescribe the physical world more accurately, and certainly not to describe it more fully. What they chiefly aimed to do was to remodel language in a more enlightened way so as to get rid of human agents and – more particularly – to get rid of souls. This topic concerned them specially because they were campaigning atheists. Their definition was above all an anti-religious campaign. It was meant as part of a much wider drive on behalf of the Enlightenment, something that clearly had plenty of point in early-twentieth-century Vienna, which was then still the church-centred capital of the Holy Roman Empire.

In this context, it is interesting to notice some more recent changes. According to Google's Ngram database, the word 'soul' 'experienced a steady decline in usage across the twentieth century, [but] this trajectory surprisingly reversed in the mid-1980s, to such an extent that as of the most recent data available, it was more common in English than the word "brain".[1] This recent reversal and the current wider use of the word have clearly arisen from the world of music, but they are by no means confined to it. As Eric Austin Lee and Samuel Kimbriel say in the introduction to their lively book *The Resounding Soul: Reflections on the Metaphysics and Vivacity of the Human Person*:

There remains something perplexing about the lingering need to express certain aspects of lived experience with reference to a 'dead hypothesis'. Why is the word soul still so damn useful? ... In historical terms, the idea of humans as 'ensouled' arose not as some bloodless hypothesis but from within a powerful set of practices concerned with fostering human potentiality and vitality. To understand this point is also to understand why, despite protestations to the contrary from certain academic circles, the soul is still very near at hand.[2]

Thus William Desmond, describing Dusty Springfield's singing, writes:

It is not the self of the singer that communicates the spread of the thrill – it is soul ... The voice raises the banal to the exalted, making more of the less, transforming, ensouling ... This is a song of ... astonishment before the being of the beloved ... Even in an age of science and technology, these ways of speaking persist. People will not forfeit their souls easily.[3]

In fact, as Austin Lee and Kimbriel remark:

There is something curious about the frequency with which the term 'soul' is now used in English in reference not to human beings but to inanimate objects. There is soul food and soulless fast food, soul music and soulful music, and apparently, according to a friend, even my old Raleigh three-speed bike has 'got soul'.[4]

Similarly, I have seen an advertisement for 'Soul Style India Pale Ale'. We may be tempted to dismiss this whole orgy of soul-talk as meaningless, but I don't think we can do that, partly because its sheer volume surely shows that it means something, and partly because the clear use of words like 'soulless' shows us exactly what that something is. This is talk about activity, vivacity, vitality, zing, and it connects them with the spontaneity of the human subject, returning that subject to the centre of the stage, refusing to accept the reductive, disinfecting campaign of scientistic puritans such as the Vienna Circle.

The awkward persistence of humans

Yet, more than a century after that circle's depersonalising efforts, this last ambition is evidently still seen as very important. Thus Steven Pinker, called on to discuss artificial intelligence, raises his flag against human agents at once like this:

Thomas Hobbes's pithy equation of reasoning as 'nothing but reckoning' is one of the great ideas in human history. The notion that rationality can be accomplished by the physical process of calculation was vindicated in the twentieth century by Alan Turing's thesis that simple machines can implement any computable function ... This is a great idea for two reasons. First, it completes a naturalistic understanding of the universe, exorcising occult souls, spirits and ghosts in the machine ... Second, it opens the door to artificial intelligence – to machines that think ... A human-made information processor could, in principle, duplicate and exceed the powers of the human mind.[5]

But here things have evidently begun to go wrong. It is quite true that Hobbes said that all reasoning is only reckoning[6] – indeed, in one of his wilder fits of reductivism he went beyond that and said that it is all only *addition* and *subtraction*. He intended this highly implausible ruling to prove that any talk not reducible to these mathematical terms is talk such as 'Accidents of Bread in Cheese, or Immaterial Substance',[7] and has no meaning.

But, however laudable that aim might be, Hobbes's general remarks about reasoning can't be defended. The term 'reasoning' obviously covers a vast range of activities from pondering, brooding, speculating, comparing, contemplating, defining, enquiring, meditating, wondering, arguing and doubting to proposing, suggesting and so forth – activities without which none of the secure rational conclusions that are being sought could ever be reached. On top of this, and still more obviously, Hobbes's remarks were never meant to refer to non-human calculators such as Turing machines. Much though Hobbes might have liked these, in his day these things had not even been thought of.

In fact, Pinker's claims make plain how obscure the meaning of the Vienna Circle's supposedly clear term 'physicalism' actually is, and indeed, beyond that, they raise serious questions about the whole meaning of reduction. What does it mean to say that reality is all reducible to certain kinds of physical entities? No doubt a rabbit, if boiled down, will be found to consist of certain kinds of physical entities. But that does not prove that, before the boiling, it was not alive or was not a real rabbit, and certainly not that rabbits are an illusion. The chemicals that the boiling process reveals are not in any way more real than the original whole animal. Indeed – unless we mean to go back to Plato's conception of the superior reality of the Forms? – the whole idea of different degrees of reality makes little sense.

Intelligibility and reality

Is the chemical and physical story perhaps more intelligible, more explanatory than the biological one? If it seems so, this surely only means that it can be fitted into a neater, simpler, more abstract conceptual scheme than an ordinary biological explanation – a scheme nearer to that North Pole of abstraction, pure mathematics. This process is often called 'explaining' and the word is used as if it meant 'explaining completely'. Thus, as we have seen in Chapter 15, Crick tells us that 'the scientific belief is that our minds – the behaviour of our brains – can be explained by the interaction of nerve cells (and other cells) and the molecules associated with them'.[8] He takes this to mean that this process is a complete account of its nature; nothing more need be said. Apparently, things not mentioned in this story are unreal, so our minds can be dismissed as illusions.

But why on earth should this follow? To *explain* something is simply to answer whatever questions arise about it, and normally there are many such kinds of question. There is nothing to stop minds, or rabbits, being understood in more than one way – fitted into more than one conceptual scheme, and still remaining as real, and as complex, as they were in the first place. In fact, explanation usually asks for a wider background, within which both ways of thinking make sense.

21

The cult of impersonality

Soul-phobia

The philosophical scientists of the Vienna Circle were, as I say, not trying to redescribe the physical world more accurately, and certainly not trying to describe it more fully. Nor indeed were they trying to redefine reality itself. They were merely trying to 'reduce' everyday speech to abstract technical terms – terms less tainted with unwelcome social ideas from everyday existence, such as consciousness and indeed life itself. Carnap's claim that all science can be expressed in the language of physics implied that the proper scientific description of (say) an apple would be a phrase translated first into chemical terms, and from them into still more abstract physical ones. But obviously attempts at this reduction would fail to convey the necessary background concepts of trees and fruit, of ripening and being eaten, without which apples cannot be explained at all. It would not succeed in describing them and most likely could not effectively describe anything.

Physics, in fact, has – like its parent, dualism – no appropriate way of describing living things, indeed, no notion of life itself. That is why the word 'biology' and the sciences that it describes have been invented, and it is why the Vienna Circle's plan of describing reality solely in the language of physics is not really practical at all. That plan is part of a vast, unworkable project of disinfection – a flight from everyday life – to which Pinker points when he writes that his little piece of Hobbesian reduction 'completes a naturalistic understanding of the universe, exorcising occult souls, spirits and ghosts in the machine'.[1] The belief that this understanding has actually been completed, and is now a secure part of science, is part of the 'modern' reductivist creed that we are discussing, the creed that so many people today urge us to believe in. Marilynne Robinson, in her shrewdly titled book *Absence of Mind*, astutely calls it the 'myth of the threshold'. As she says, in what we call modern times this conditioning has been part of our upbringing:

> I was educated to believe that a threshold had indeed been crossed in the collective intellectual experience, that we had entered a realm called 'modern thought' and we must naturalize ourselves to it … [The details of this crucial event

were not made clear and were not thought to matter] What we had learned from Darwin, Marx, Freud and others were insights into reality so deep as to be ahistorical …

The schools of thought that support the modernist consensus are profoundly incompatible with one another, so incompatible that they cannot collectively be taken to support one grand conclusion … [And yet there is still a] core assumption that remains unchallenged and unquestioned … [It] is that the experience and testimony of the individual mind is to be explained away, excluded from consideration … The great new truth into which modernity has delivered us is generally assumed to be that the given world is the creature of accident … Once it was asserted and now it is taken to have been proved that the God of traditional Western religion does not exist … An emptiness is thought to have entered human experience with the recognition that an understanding of the physical world can develop and accelerate through disciplines of reasoning for which God is not a given.[2]

Pinker's casual, super-confident reference to this orthodoxy is typical of our times except that he attributes it to a philosopher rather than a scientist. The creed is usually taken to be essentially scientific, established by some technical discovery or other – probably by the discovery that the earth is not the centre of the solar system, though this is not always distinguished from the discovery that the earth is not flat. Thus, this contemporary orthodoxy is well known to date back at least to the sixteenth century, and it can only be called 'modern' by contrast to that century's best-known predecessor, the 'Middle Ages'.

As Robinson says, however, this creed has also been linked with many later sages – Darwin, Marx, Freud and whatever other notables are viewed at any time as able to explode an old-fashioned, supposedly religious approach. Its main function, however, is not to get rid of God but to support materialism by eliminating human subjects. As Pinker says, it exorcises 'occult souls, spirits and ghosts in the machine', in order to make room for abstract science.

Constructing the vacuum

Thus, instead of getting rid of dualism, as was intended, this 'modern' outlook keeps to a strange, one-legged variety of it which simply establishes physical science as the champion of old-fashioned Matter against old-fashioned Spirit. What this destroys is not superstition but the idea of the individual thinker. It aims to undermine the observers' and theorists' own minds, their confidence in their own judgement, in themselves as able to judge. As we have seen, it leads to Crick's vacuum, where minds are held to have vanished altogether.

This sweepingness can, of course, look good on the title page. But it has increasingly awkward consequences when – as is bound to happen – reasoners need to take notice of their own impressions, their opinions, their convictions, their doubts, their range of consciousness, their relations with others, their experiences and their inner life. Serious physicalists must then claim that these items – which they are actually using – don't exist. ('I myself,' says Crick, 'like many scientists, believe that the soul is imaginary and that what we call our minds is simply a way of talking about the functions of our brains' (RB p.3, note). So who is this 'I myself … '?) Whether we call them illusory, like Crick, or simply refuse to talk about them at all on the grounds that they are vulgar everyday matters, we are no longer supposed to mention these things, which have been excluded from the province of science. But since we all are vulgar, everyday people, needing to talk about vulgar, everyday things all the time, even when we are doing science, this is very inconvenient. As Merleau-Ponty explains:

> I cannot conceive myself as nothing but a bit of the world, a mere object of biological, psychological or sociological investigation. I cannot shut myself up within the realm of science. All my knowledge of the world, even my scientific knowledge, is gained from my own particular point of view, or from some experience of the world without which the

symbols of science would be meaningless. The whole universe of science is built upon the world as directly experienced, and if we want to subject science itself to rigorous scrutiny and arrive at a precise assessment of its meaning and scope, we must begin by reawakening the basic experience of the world, of which science is the second-order expression. (2015, 10–11)

The area where this need is most striking is, of course, experience itself, which, in spite of being subjective, often needs to be described, and notoriously contains, as Merleau-Ponty says, the whole mass of solid empirical evidence on which science depends. Thus, the case for the existence of rabbits all rests eventually on accepting particular people's reports of having personally seen, felt, heard, touched or tasted particular rabbits – reports which inevitably contain terms quite alien to the language of physics. Because of this, Pinker's 'naturalistic understanding of the universe' can only be called 'naturalistic' if *nature* – the real universe – is understood, for this purpose, as simply consisting of the abstract world that physics describes. Only what is mentioned by physics is then real, and the physicist can safely say, in the words of the old rhyme:

I'm the Master of this College,
What I don't know isn't knowledge.[3]

Time trouble

This strange idea has been remarkably influential, though scientists often don't notice that they are committed to it. It is surely the background of the still stranger current view, officially accepted by many physicists including Einstein, that time itself is not real. As Tim Maudlin says:

> Some physicists are very adamant about wanting to say things about [time]. Sean Carroll for example is very adamant about saying that time is real. You have others saying that time is just an illusion, that there isn't really a direction of time and so forth. I myself think that all the reasons that lead people to say things like that have very little merit, and that people have just been misled, largely by *mistaking the mathematics they use to describe reality for reality itself.* If you think that mathematical objects are not in time, and that mathematical objects don't change – which is perfectly true – and then you're always using mathematical objects to describe the world, you could easily fall into the idea that the world itself doesn't change, because your representations of it don't.[4] (Emphasis mine)

In short, you come to believe your own propaganda. You forget that there is actually a world beyond the abstract, timeless mathematical subject-matter on which you work. You forget this all the more easily because many basic doctrines of physics really are not concerned with time and change. They simply abstract from it. So physicists studying these matters can go on as if they are indeed dealing with a static, unchanging world. When they do this, their brain's left hemisphere, which always applauds narrowness and hates to be reminded that the wider world exists, cheers and congratulates them. Then they can burn incense to their patron saint Pythagoras, who originally launched the idea that All Is Number, and who never got around to saying, 'Well, of course, you see, only in a way ... '

Thus the doctrine that time is unreal works rather as if an anatomist, being entirely occupied with skeletons, came to think that bones are the only reality and the soft parts surrounding them are all illusory. But the illusion that time is unreal has been found more convincing than the skeleton story would be because of its history. Modern physics still carries a special authority, inherited from its predecessor – the material half of dualism – which had great prestige, and which did indeed rule clearly that all real matter was inert.

Questions, however, still arise about just what this story about time's unreality can mean. When we say that something is not real we are always saying something positive about it. We are saying either that it is imaginary, or a pretence, or a mistake, or an illusion, a lie or a fake, or a hallucination – in fact, that it is not continuous with the real world around us and need not be taken seriously.

Now if the physicist, while working on this topic, is interrupted by an alarm clock which tells him that the moment for a most drastic change in his life has now struck, and if he then responds by merely muttering, 'Oh well, luckily time is not real,' then we can conclude that he interprets this principle in one of these ways, and we can ask him which way it is. But if he responds instead in one of the more active ways in which people usually respond to alarm clocks, then the thing is simpler. He just doesn't believe what he is saying.

22

Matter and reality

More Soul-Phobia: Behaviourism

Thus physicalism, in its strict Vienna Circle interpretation, really can't be combined with any plausible belief in the commonsense world around us, nor therefore with any plausible belief in ourselves. We know, of course, that people do manage to combine sets of beliefs that don't fit well together, and we put up with this so long as no-one lays them under our nose. But physicalists do lay this one under our noses because they express their point so strongly and paradoxically. Indeed, even old-fashioned materialism already does this. The point of claiming that 'nothing exists except matter and its movements and modifications' is not just to get rid of God. It is primarily to get rid of minds, to show that any reports of effective mental activity – including our own – must be false or meaningless. This idea does not just degrade subjectivity; it blots it out entirely, inviting us into Crick's vacuum. As Robinson remarks:

> The thing lost in this kind of thinking … is the self, the solitary, perceiving and interpreting locus of anything that can be called experience. It may have been perverse of destiny to array perceptions across billions of subjectivities, but the fact is central to human life and language and culture, and no philosophy or cognitive science should be allowed to evade it.[1]

In short, subjectivity is not just a term of abuse. It is an objective fact about the world. It is true that we each do our thinking separately and subjectively. Our imaginations have to work hard to bring us together sufficiently to believe in a shared world, which we can then see as *intersubjective*, or *objective*.

When people complain during an argument that their opponents are being *subjective*, they are not referring to this well-known general fact of separateness. They are saying that some particular private consideration has intruded into the discussion on a topic where it was not relevant. And the essential point here is not the subjectivity; it is the irrelevance. If the discussion is about rabbits, and someone supports his argument by mentioning a rabbit that he saw last week, it is not in order to silence him by saying, 'But that is subjective. It is merely anecdotal folk-psychology. We must wait till this point has been reported in a peer-reviewed journal.' Unless we have reason to distrust the speaker, we have to accept that what he reports has been an objective fact like any other, and in the end it will have to be accounted for.

Constructing the vacuum

How, then, can so central a thing as this crucial human subject have got lost? It got lost because the fashion of the day provided no usable way of talking about it, and in recent times, this has meant essentially because of behaviourism. As I complained long ago in *The Myths We Live By*, the pre-existing view of Enlightenment rationalism about our Selves had long been badly flawed. It showed the essential Self as consisting in *reason*:

an isolated will, guided by an intelligence, arbitrarily connected to a rather unsatisfactory array of feelings, and lodged, by chance, in an equally unsatisfactory human body. Externally, this being stood alone. Each individual's relation to all others was optional, to be arranged at will by contract. It depended on the calculations of the intellect about self-interest and on views of that interest freely chosen by the will.[2]

Behaviourism formalized this picture of the human condition and, for a time, made it compulsory throughout the social sciences. It is an implausible picture, which is not surprising because it is essentially political. It was invented largely for particular, quite urgent, purposes connected with civic freedom and the vote. The social-contract conceptual scheme was a tool, a wire-cutter for freeing us from mistaken allegiance to kings, churches and customs. And in general the notion was, at first, carefully not used in places that did not suit those purposes.

As its influence spread, however, it did get used in other contexts, with various dubious effects. One of the most obvious of these emerged in a strong connection with gender. Social-contract thinking had originally been applied only to men, and any later attempts to extend it to women aroused painful indignation and confusion. Each man – each voter – was conceived as a unit, representing and defending his house. There was no question of its other members needing to speak for themselves, nor of one voter needing to understand the others. It is no trifling matter that the whole idea of an independent, enquiring, choosing individual, an idea central to Western thought, has always been essentially the – somewhat romanticized – idea of a solitary male. No wonder that, at the time, the public gave such a welcome to Robinson Crusoe.

As I say, this one-sided political language clearly never provided a realistic way of talking about the range of problems that actually beset human life. It gives no expression to our natural sociability and friendly co-operation, our constant need to interact with others and yet also sometimes to oppose them, nor, what is quite as important, to our equally constant need to deal with our own and each other's inner conflicts. It suggests that the work of the arbitrator, Reason, is entirely intellectual, a matter of calculating possible profits and losses, and of preventing the unenlightened passions from interfering with the rational results of this arithmetic.

23

The mystique of scientism

Imaginative visions

This all bears on something important which we have considered earlier, namely, the myths, the dreams, the fables which distort our view of contemporary science. The awe with which we tend to view current research (about which we actually know very little) does not rest only on its real merits. It also flows from an impression that this kind of learning is something grandly *impersonal*, securely free from individual influences, finally and permanently true. Even though we know that in the past scientific orthodoxy has sometimes been found to be mistaken – for instance, about phrenology, or the effectiveness of bleeding as a medical procedure – this doesn't affect our respect for its successors today. Instead, it only deepens our reverence for modern medicine, which we think has corrected all these errors.

As I have been suggesting, the idea of physical science now carries a strong and effective symbolic weight. Something called 'science' has succeeded to the position of authority in our culture which used to be held by religious creeds. People use the word 'materialist' just as some of them used to use the word 'Christian', as a synonym for 'sane, normal, rational person'. It is now the central example of a compulsory doctrine, something that has to be believed, whether you understand it or not. It counts as an oracle. And this status has disturbing effects on the way in which we think about the details of science itself. Instead of seeing the physical sciences as real, but limited, sources of knowledge about material facts, scientism now calls on us to revere them as the metaphysical source of all our knowledge. Physics and chemistry no longer appear just as two stars among many, but as parts of a super-sun, the final form of knowledge, a terminus for which all other forms of thought are only provisional sketches.

One intriguing angle on all this concerns a shift in the content of superstition. As I've mentioned, scientistic orthodoxy prides itself on having demolished superstition by removing religion, which is supposed to be unable to survive in the current rational atmosphere. It is interesting to see how promptly this move was followed by the rise of new superstitions designed to support scientism itself. Let us look at two examples.

One is Professor Krauss's claim that science provides the ultimate basis for reality – the ultimate metaphysics. Is this claim superstitious? It does not, of course, refer to any religion, but then, as the dictionary makes clear, superstition does not have to be religious; it can mean unfounded belief in general. Explaining this, the OED quotes, very appropriately, from no less a physicist than James Hutton the remark that, 'I am afraid there are many men of science that only believe the theory of heat and cold in prejudice or superstition, i.e. without having seen its evidence.'

That is presumably how Krauss has come to suppose that his favourite 'metaphysics' – physicalism – was provided by science, when in fact, as we have seen, it grew out of conceptual doctrines that had long been brewing in the kitchens of European dualism, and was given its final form by the philosophers of the Vienna Circle.

My other example is, I think, an even more unmistakeable case of pure superstition. It is this belief in the 'singularity' itself, the supposed transition expected to occur at the point where artificial intelligence finally outdoes the human variety and takes over from it. This is not supposed to be just a fairy-story. It is meant as a factual prediction, comparable to a conviction that King Arthur will indeed return one day, or that the world will come to an end next April. And of course it has no more solid evidence than those predictions.

This singularity started life on an honest basis, as a theme in science fiction, a respectable genre which, from Wells's days on, has been clearly understood to be an imaginative art, valuable for the indirect light that it sheds on the real world, but absolutely not a cheap competitor in the field of true history. I have noticed that this topic of artificial intelligence has for some time seemed to be slipping out over the borders of that harmless zoo. So, wanting to see what was going on, I picked up a book called *What to Think About Machines That Think*,[1] a collection of some two hundred short essays about artificial intelligence by all sorts of highly qualified people, including some distinguished scientists. So – though I thought its title was rather credulous – I started looking through it.

What I hoped for and expected to find, among other things, was the sort of investigation that is now so generously given to all sorts of imaginative literature, including myths and fairy-stories. I expected an enquiry into the meaning of artificial-intelligence talk, its sources and motives, its function, its public, its appeal, the elements of truth that there might be in it, the uses that are being made of it and the harm that it might do. I looked forward to all that, because I thought it would be very useful. But I could not have been more mistaken. Far from this, many of these experts apparently take the outlines of this story as established fact and occupy themselves only with details. Though there are some sceptical voices, and many who don't want to be involved,

I found only one real protest about the whole project. This is 'The Singularity, An Urban Legend?', a sharp piece by Daniel Dennett, which, by contrast, takes that challenge seriously.

Dennett asks, are these questions about artificial intelligence perhaps actually important? Ought we to pay serious attention to these machines because of the dangers that they may pose in the future? And he answers, No:

I think, on the contrary that these alarm-calls distract us from a more pressing problem, an impending disaster that won't need any help from Moore's Law or further breakthroughs in theory to reach its much closer tipping-point. After centuries of hard-won understanding of nature that now permits us, for the first time in history, to control many aspects of our destinies, *we're on the verge of abdicating this control to artificial agents that can't think, prematurely putting civilization on autopilot* [emphasis mine]. The process is insidious, because each step of it makes good local sense, is an offer you can't refuse … To lay people, AI means passing the Turing test, being humanoid … But the public will persist in imagining that any black box that can do *that* (whatever the latest AI accomplishment is) must be an intelligent agent much like a human being, when in fact what's inside the box is a bizarrely truncated, two-dimensional fabric that gains its power precisely by not adding the overhead of a human mind with all its distractability,

worries, emotional commitments, memories, allegiances …
It's not a humanoid robot at all but a mindless slave, the latest
advance in autopilots. [2] (Emphasis mine)

In short, this whole controversy is something worse than a
waste of time. It is a damaging self-deception. What is needed
now is sharp attention – human minds determined to direct
their painful efforts to a most difficult set of problems, to
penetrating and shifting a dangerous contemporary delusion.
All this speculation about the details of new, expensive gadgets,
as if they could take over the work of reshaping civilization from
us, is a mere distraction.

It is rather remarkable that this protest of Dennett's seems to
be the only place in this book where its authors attend in any way
to concern about our planet's actual future prospects, apart, of
course, from the proposed development of artificial intelligence
with which they are obsessed. They ask no questions about the
terrestrial background – climatic, geological, economic, social,
political or whatever – that can be expected to shape future
conditions on earth, conditions which would certainly have to
be exactly right to make their enterprise at all possible. The only
prudential consideration that seems to bother them is something
immensely distant – the need to time space-colonization so as to
avoid being caught out by the remote future activities of the sun
and the rest of the universe.

The best way to illustrate this pervasive attitude is, I think, to quote fully from the prophecies of a single highly distinguished contributor, one who does not differ from the rest in any way except in being exceptionally eminent:

ORGANIC INTELLIGENCE HAS NO LONG-TERM FUTURE

MARTIN REES

Former president of the Royal Society; emeritus professor of cosmology and astrophysics, University of Cambridge; fellow, Trinity College; author of *From Here To Infinity*

... [A]ssessments differ with regard to the rate of travel, not the direction of travel. Few doubt that *machines will surpass more and more of our distinctively human capabilities, or enhance them via cyborg technology* ... In a long-term evolutionary perspective, humans and all they've thought will be just a transient and primitive precursor of the deeper cogitations of a machine-dominated culture extending far into the future and spreading far beyond our earth ... It's not hard to envisage a hypercomputer achieving oracular powers that could *offer its controller dominance of international finance and strategy* – this seems only a quantitative (not qualitative) step beyond what 'quant' hedge-funds do to-day ... But once robots observe and interpret their environment as adeptly as we do, they will truly be *perceived as intelligent beings* to which (or to whom) we can relate – at least in some respects, *as we relate to other people* … But what if a hypercomputer developed a mind of its own? If it could infiltrate the Internet – and the 'Internet of Things' – it *could manipulate the rest of the world*. It may have goals utterly orthogonal to human wishes – or even treat humans as an encumbrance. Or (to be more optimistic) humans may transcend biology by *merging with computers*, maybe subsuming their individuality into a common consciousness …

So what about the post-human era, stretching billions of years ahead? … The potential for further development [of quantum computers] could be as dramatic as the evolution from monocellular organisms to humans. So, by any definition of *thinking* [emphasis original], the amount done by human-type brains (and its intensity) will be swamped by the cerebrations of AI … Interplanetary and interstellar space will be the preferred arena, where robotic fabricators will have the grandest scope for production and where non-biological 'brains' may develop insights as far beyond our imaginings as string theory is for a mouse …

So it won't be the minds of humans, but those of machines, that will most fully *understand the world*. And it will be the actions of autonomous machines that will *most drastically change the world* – and perhaps what lies beyond.[3] (Emphasis mine, unless stated otherwise)

Is this extract enough to show why I say that new superstitions are being invented to replace the old religious ones? Here, without any empirical evidence at all and without any moral background, a purely calculative heaven has been arbitrarily invented, a heaven with its own population of new gods – gods who can apparently be trusted, though we know nothing at all about their minds and wishes – trusted, that is, to carry on throughout the universe exactly the same sort of intellectual activities that we (or some of us) now greatly admire, and to bring those activities to a triumphant conclusion. In fact, they can be relied on to be just like ourselves, only much cleverer and more successful.

Is all this perhaps just a harmless boast, not meant as serious science? It could be, but if so people need to be plainly told so. Richard Dawkins already blurred this important frontier when he suggested in the preface to *The Selfish Gene* that his readers should enjoy that book simply as an exciting mystery story. He was trying to eat the fiction-cake and also have it – to produce extra enjoyment without admitting that the book, which indeed contains a great deal of fantasy, might not actually all be literally true. But without that explanation, prophecies like those we are now considering are naturally taken as fable or legend rather than as science.

They are naturally so taken, not only because of their lack of supporting evidence but also because their sponsors themselves so clearly don't believe their own message. If they did, they would surely, by now, not be writing books but busy with frenzied activities trying to make suitable preparations for the alarming change that threatens them. This drastic revolution is, after all, not just something that will occur many parsecs off in the distant future. It is expected here quite soon; indeed, its chief contemporary champion, Ray Kurzweil, has named 2045 as a plausible date. That, therefore, is presumably the point at which the machines, having outgrown their present advisory role in human affairs, can be expected to begin taking over (as Rees puts it) total 'dominance of international finance and strategy'. And since that dominance will surely include the running of universities and similar intellectual institutions, places in which the artificial-intelligence experts themselves mostly work, they should expect their employment conditions to change dramatically quite soon.

Rees does indeed try to soften this prophecy by saying that this kind of machine will offer that political dominance to 'its controller'. But what makes him think so? And who is the controller? He explains that these robots will be intelligent beings, our social equals, to whom we can normally relate as we do to other people. This has to mean that we can disagree with them, and can expect those disagreements to be settled, like ones between human colleagues, by argument or (failing that) by force of personality. Rees asks what would happen if one of the machines 'developed a mind of its own', as if that were rather an unlikely hypothesis. But since they are supposed to be cleverer than we are and to be getting still cleverer all the time, there seems nothing in the least unlikely about this, nor about their often winning arguments.

In short, the picture that Rees offers of super-intelligent but mindless and super-docile servants does not make sense. It seems to be rather like the mistake that rich climbers, ambitious to 'conquer' Everest, recently made when they assumed that the Sherpas who helped them – and who themselves were climbing Everest all the time, unreported by the news, simply as part of their job – had no views of their own and would always accept their orders.

This, therefore, is one of many topics on which an enquirer must point out, however regretfully, that the whole story is not just science fiction but bad science fiction. A good SF writer (such as Isaac Asimov, who paid a lot of attention to these matters) would surely have seen these clashing demands within the plot in advance and would have done what he could to sort them out before completing his account. To do this, however, he would have had to make drastic changes in the shape of the whole story he was telling. It is not possible both to credit these new machines with ever-increasing cleverness and to leave the human race in its accustomed secure position as lords of the earth and the summit of evolution.

Rees does indeed mention a solution for this difficulty, a solution which is laudably ambitious, but is unfortunately quite unintelligible. It is that 'humans may transcend biology by merging with computers, maybe subsuming their individuality into a 'common consciousness'. But to make sense of this you have to be able to imagine it – to have some coherent picture of how it would actually work. Does this merged creature (for instance) talk, or does it have a keyboard? And what would a halfway stage in the merging – a partially computerized human – look like? Since we no longer have with us either Asimov or some yet wilier exponent of his art who could give a sense to this suggestion, it remains just a vacuous string of words.

The idea clearly is that (again) we ought somehow to be able to eat this cake and have it, that, by changing ourselves in this way, we should be able to get the extra calculative powers of the machines without paying the price of altering our own natures. This transformation is expected not just to solve our current problems, but to make us (or our successors) capable of further amazing intellectual feats in the future.

Why, however, should anyone expect these extra calculative powers to make the difference that is needed? The confusions that now afflict human life are not due primarily to lack of cleverness but to ordinary human causes such as greed, bias, folly, meanness, ignorance, ill-temper, lack of common-sense, lack of interest, lack of public feeling, lack of teamwork, lack of experience, lack of conscience, perhaps most of all to mere general lack of thought. In our search for this imagined extra cleverness we are (as Dennett pointed out) hoping to employ mindless slaves, beings who would only do what we tell them and would never mention any topic that we do not give them as part of their task; this would mean that they could not in any way out-think us. And – as Butler noted (see Chapter 19) – in cultivating them we are distorting our own natures by attending to the working of these tools rather than directly using our own powers to solve the problems that threaten us, problems such as chronic human narrowness and – at least equally with that – climate change.

24

The strange world-picture

Letting in the mind

I should now come back from these rather fantastic realms and explain what is my central point: what has actually made me write this little book?

What makes me write books is usually exasperation, and this time it was a rather general exasperation against the whole reductive, scientistic, mechanistic, fantasy-ridden creed which still constantly distorts the world-view of our age. That creed, though often discredited during the last century, still bears the flattering name of the 'modern' outlook. In fact, it still commands so much respect that new suggestions can be justified merely by being hailed as the latest thing. It is revered as the only possible alternative to a supposedly unthinking, moralistic and religious previous orthodoxy. Thus, in a time of rapid and continuous change, our official creed still spins round a circle of possible responses to the supposed beliefs of our late-Victorian predecessors.

Plenty of people have, of course, already complained about this. Most notably, in 2012, Tom Nagel published a little book which he called, with his usual clarity, *Mind and Cosmos: Why the Materialist Neo-Darwinian Conception of Nature Is Almost Certainly False*. That book was not actually the first chirrup of the dawn chorus. But it was a good loud one, and what made it stand out – what made those sleepers who particularly disliked being woken so early pull up extra bed-clothes against it – was that it was remarkably comprehensive. It advised that the whole concept of Cosmos should be so radically expanded that it could include not only Life but Mind:

> The great advances in the physical and biological sciences were made possible by excluding the mind from the physical world ... But at some point it will be necessary to make a new start on a more comprehensive understanding that includes the mind ... With the acceptance of the big bang, cosmology has also become a historical science. Mind, as a development of life, must be included as the most recent stage of this long cosmological history, and its appearance, I believe, casts its shadow back over the entire process ... My guiding conviction

is that mind is not just an afterthought or an accident or an add-on, but a basic aspect of nature … [This, he says, is not for religious reasons.] I do not find theism any more credible than materialism as a comprehensive world-view … These two radically opposed conceptions of ultimate intelligibility cannot exhaust the possibilities.[1]

And this startlingly wider horizon should mean that it becomes possible to use again the recently vetoed concepts of purpose:

The idea of teleological laws is coherent, and quite different from the idea of explanation by the intentions of a purposive being who produces the means to his end by choice … In spite of the exclusion of teleology from contemporary science, it certainly shouldn't be ruled out *a priori*.[2]

I don't think anybody had pointed out before this quite how completely the 'modern' scientistic outlook has depended on assuming an absence of mind – depended, in fact, on theorists pretending that they themselves weren't really there creating it. Even before Crick started denying our existence they have long gone to great lengths to avoid mentioning even the most obvious truths about our subjectivity. This resolute soul-blindness raises a host of difficulties, some of which I have tried to examine in this book.

The inner world is real

The area of this history that I have wanted to investigate most fully is the imaginative aspect of the matter. I wanted to look at the myths, fables, images, fantasies, dreams and carefully constructed world-pictures by which this scientistic outlook has been built up and sold to the wider public, often with a label saying either 'science' or 'philosophy' or both. I have long suspected that a central cause of this trouble was the increasing specialization of our age – the growing tendency of educators to supply more and more separate examinable qualifications for everything, rather than putting things together intelligibly. In particular, the idea that philosophy, like pharmacy, is just one more of these neat detachable skills – a knack needing only standard courses in applied logic done under a qualified teacher – is surely particularly misleading. This approach is as anti-philosophical as it is anti-scientific.

Official teachings of this kind are not, of course, a necessary part of philosophizing. Many great philosophers of the past – Epicurus, Hobbes, Hume – scarcely even attended such courses, still less followed their teachers. They were primarily trying, on their own account, to bring together aspects of life which they saw had become separated, or to disentangle ones which had got dangerously entangled. They saw the twists and turns that were already distorting human thought – and with it human life – in perspective. They stood back from immediate problems and asked how to straighten the whole background so as to put these twists right.

This often involved asking large and unexpected questions, which really don't have easily summarized answers. That is why in this book I keep coming back to a paradox. On the one hand, I want to emphasize that there really *is* only one world, but also – on the other – that this world is so complex, so various that we need dozens of distinct thought-patterns to understand it. We can`t reduce all these ways of thinking to any single model. Instead, we have to use all our philosophical tools to bring these distinct kinds of thought together. Most of these kinds do have their own uses and several can often be used together. In fact, as Kipling explained:

> There are nine and sixty ways
> Of constructing tribal lays
>   And every single one of them is right.[3]

The questions that puzzle us may be, for instance, scientific, historical, practical, poetical, moral, social, logical, political or religious, and they must all be discussed in their own appropriate terms. Yet we must also find the connections between them. And the tidy specialization of today's academic world can make this seem almost impossible.

So our main philosophical need is not, as many eager arguers think, to orchestrate and embitter the conflicts between them so as to settle them finally. It is to find ways of bringing the two sides together. As Hegel put it, we should keep finding ways of combining the best parts of a thesis with its antithesis to make a new synthesis.

In trying to do that, however, we are torn between our thirst for agreement, for unity, and our deep dislike of some of the ideas involved. So we fight to keep these out of that final agreement. These are the conflicts, the divisions in our basic attitudes out of which philosophical disputes have always grown, and on which they still constantly estrange us today.

Insides and outsides

In order to illuminate this general aim I have kept coming back to two particularly clear examples of these reconciliation problems – the meaning of mental health, especially of mental illness, and free will. Mental illness obviously has two interdependent aspects – the impersonal, medical angle and the patient's own viewpoint. We try to balance these fairly. But this is hard, both because of particular details and because of our general bias for or against some wider aspect of science or humanity.

Treating the two sides together involves looking in at a single scene – as we might do at an aquarium – through two separate windows which reveal different views. Or again, it is like the skill we use when we study the first pages of our atlases and try to bring together the different maps that appear there – physical, political and barometric charts of the same piece of territory. This sort of comparison calls for a kind of controlled mental squint, a reconciling, two-sided apprehension of a kind which, as I have been suggesting, is really difficult and is the peculiar province of philosophy. It often shows the need for serious background thinking, both about the facts and about the various ideals involved.

About mental illness there has surely been some real progress during the last century, because people are growing more aware of the problems and more willing to work on them. That, however, has certainly not happened about my other chief example, the problem of free will. Choice too is a condition where we see what is happening in two quite distinct ways – as actors and as spectators. But, instead of noticing that these methods need to be brought together, theorists in our tradition have lately preferred simply to ignore the inner data altogether as being subjective. Thus, people who want to be thought scientific now often simply follow their left hemispheres and proclaim a materialist orthodoxy, which treats human choice as merely an outward event in the physical world, something which cannot be influenced by thought because it is fully shaped by the movements of the brain cells.

Yet there is a strange gap here between theory and practice. The next time these people themselves have to make a choice – for instance, even a small choice in finding the right words to use in writing an article – they do not, as their theory demands, simply sit back and wait for their brain cells to take the decisions for them. (We have all tried doing that and we know it doesn't work.) Instead, materialists, just like the rest of us, have to *attend to what they are doing*.

For instance, it is clear that that distinguished behaviourist prophet B.F. Skinner had to think quite hard in order to produce the ingenious excuses that he always gave for his implausible doctrines. Brain cells alone, without help from an attentive human, could not possibly have done this for him. In fact, materialists, like anybody else, need to think, and it is their thoughts that eventually determine the words that they write.

There is no inert matter

Minds, then, are effective as well as bodies. This means that we should now, for a change, think carefully ourselves about the relation between these two quasi-elements. Indeed, more than that, we should at last ask seriously why it was ever thought necessary to divide the world into two quasi-elements in the first place. This is not, of course, a question about the division between our inner and the outer viewpoints, which is a real division. The unreal part of the story is the traditional doctrine about *what* these two things are, especially the notion of matter as a dead, inert mass made up randomly of little disconnected particles.

This is the picture that made T.H. Huxley, and indeed his successors who talk about the 'hard problem' of consciousness today, so incredulous about matter's being able to generate consciousness at all. Thus, quite recently, Henry Marsh, clearly an exceptionally sensitive and sophisticated surgeon, wrote in distress of 'the extraordinary fact, which nobody can even begin to explain, that mere brute matter can give rise to consciousness and sensation'.[4]

What, however, forces us to think of matter as merely 'brute' in this way? If we accept the facts of evolution, we already know that this same 'matter' is what has given rise to the whole world of living things – that it began to do so as soon as conditions on the earth made this physically possible, and that it has continued to do it ever since by a series of subtle stages which have taken it across many surprising gaps, including indeed the gap involved in the origin of life itself. This history shows that, right from the start, this 'matter' had within it the potential for forming all these highly various and sophisticated things. It cannot therefore be the inert, inactive, neutral stuff of dualist theory.

And of course, in real science, nobody since the eighteenth century has believed in that kind of inert, neutral, dualistic matter at all. The discovery of electricity destroyed its initial plausibility, and since that discovery we have been constantly learning more and more about the spontaneous activities of molecules and the active part that their co-operation plays in the development of life. This is indeed a main occupation of present-day biology. So, from the point of view of today's science, we now do not live in an inert, helpless, static, atomized, dead, divided world but in a coherent, continuous, active one, quite capable of generating these new activities by self-organization. Accordingly, the dogmatic materialism which for some time was a compulsory orthodoxy among scientific-minded Westerners has now lost its meaning.

Wish fulfilments

Unfortunately, however, this kind of materialism still haunts us even without a meaning, because it is part of the mythical dualist world-view, and, as we have seen, such myth-based visions are strong. They do not readily yield to correction by Time and Truth. Indeed, being part of the temper of the age, they don't die simply because they have been so often debunked. Instead they are repeatedly born again within us and die – or fade and revive – in response to changes in social moods and wishes.

Thus the shift among us Westerners during the last two centuries from trust in God towards trust in a machine-based future has not actually been produced by any force of argument. It is a self-propelling mood-shift in which arguments play only a superficial part. Its mythology, which at present centres on space colonization and includes the singularity, is not derived from facts; it is a spontaneous piece of wish fulfilment. The remarkable thing here is the speed, the promptness with which new machine-based imagery keeps appearing to replace what is lost. The moonshots, for instance, were at once seized on and cited as the final, supreme example of human achievement, the pointer to all future goals, as if everything else that *Homo sapiens* has done were relatively trivial. The crude fact that these achievements led nowhere in particular – that they were only a move in the Cold War and no use, so to speak, was ever made of them – has never damaged their status as the contemporary vision of Heaven. In place of any such vulgar practical support, science fiction writers, ably backed by rocket-building tycoons, have continued to supply them with myths promising unlimited glory in the future.

This is the mechanistic background that has made scientism look plausible. It is a mythical way of exalting the physical sciences as the supreme human intellectual activity. Faith in the supremacy of these sciences over all other kinds of thinking is taken as establishing that the world actually consists entirely of something called matter, rather than people's seeing that faith is itself a mere consequence of this belief. As we have seen, the concept of matter itself is obscure, but the main point of the materialistic creed has always been its destructive side – the determination *not* to think in terms of minds or souls.

We are therefore called on either to conclude, with Crick, that we ourselves are not there – are merely imaginary – or (if we find that rather hard to understand) to avoid subjective talk altogether, translating it always into talk about brains or other material objects. This reductive approach leads, however, to such a distortion of our thought, such a twisting of all natural means of communication that it cannot really prosper.

In fact, a whole heap of considerations have gradually made it obvious that the inner, subjective point of view is every bit as natural and necessary for human thought as the outside objective one. Indeed, neither of these positions can really exist without the other. As I have previously pointed out, you can't have the outside of the teapot without the inside. The attempt to separate them – to reject all the most direct means of contact that we have with the world around us – is perverse and unworkable. As Marilynne Robinson has remarked, chronic absence of mind really is not a useful condition for everyday living.[5] So, since we find that we are here in the world, perhaps we had better accept our own presence and make the best of it. But what is the best way to use philosophy in doing that?

Conclusion

One world but many windows

The battle of behaviourism

On the first pages of this book I asked a question about our purpose in philosophizing. I enquired then what our efforts should be aiming at. And I noted that some professional philosophers now tell us that we should start by studying only the philosophy done among them during the past twenty years. But, as anyone who has watched that scene knows, professional philosophers are likely to have mainly occupied those twenty years in following up quarrels that had been developed there during the previous twenty, and so on backwards. It is largely chance what they happen to be concentrating on now. We surely need to look for a fresh start – to see what problems are really

giving trouble today. So, as I then suggested, we should look for
the main currents of thought that are active at the time, see if
there are conflicts between them – or places where they simply
fail to meet – and try to bring them together.

If we want to understand why this is necessary and why it so
often fails to happen, I think it may be worthwhile to look at some
of the cultural dramas that have been going on since I myself first
got tangled in the philosophical scene. In the 1950s, two main
influences prevailed there among the learned, behaviourism
and existentialism. Behaviourism, which was a strong offshoot
of scientism, was not supposed to be a creed, but it certainly
ruled as one. Officially, it was just the psychological – or rather
philosophical – theory that only actions were real, which meant
that psychologists were forbidden ever to mention motives,
ideals or other background features of the inner life because
these were unscientific. Actions could be caused only by other
actions.

Throughout the social studies, this rule was so sternly enforced
that, for a long time, psychologists tended to lose their jobs if they
mentioned such unwholesome topics as 'motives', 'feelings' or,
more particularly, 'consciousness'. This taboo lasted until some
time in the mid-1970s when, as behaviourism weakened, some
daring penguin (I think it was Nicolas Humphrey) boldly dived
into these dangerous waters by openly mentioning consciousness
in print. To the general surprise, the penguin survived, and after

this the 'problem of consciousness' was accepted and became everyday business.

During this behaviourist epoch, however, philosophers themselves had not been so explicitly drilled. Yet the sense that it was unprofessional ever to mention motives – vulgar and unscholarly to touch on the inner background and meaning of actions – actually became very strong among them and it has had a lasting influence on their writing. It drew heavily on the scientistic background that I have mentioned, whose moral implication was that people studying the humane subjects ought not to pollute their argument by mentioning non-scientific ideas or ideals of their own. They ought, as far as possible, always to imitate the 'hard sciences'.

As for existentialism, it too confined its attention to overt acts, but in doing this it concentrated chiefly not on science, but on a moral value – the importance of freedom, the need to generate one's own choices rather than being influenced by custom or other people's views. It therefore shared with behaviourism the conviction that 'there is no such thing as human nature'. This denial was meant to stop people from excusing their sins against freedom by claiming that these sins were only natural. And Marxism, which was another orthodoxy of the day, used this same denial to prevent people from excusing other sins – this time, sins against economic correctness – by saying that these too were 'only human nature'.

Bringing back human nature

What surprised me about all this was the strong element of overkill. It would surely have been quite easy to answer all these bad excuses without needing to say something which everyone knows is not true – namely, that human beings have no natural motives, no inborn tendencies, that they are 'blank paper at birth'. (You can't look at a live baby for long without seeing that it is not blank paper.) What's more, I found it extremely odd – indeed, unscientific – that human beings were being considered on their own as an isolated case like this, without being compared with other animals, which plainly do have such inborn behavioural tendencies.

Accordingly, in my innocent hopefulness, I wrote a book called *Beast and Man*, which was meant to show that human nature does exist, that it links us closely with the other animals, and that it does so in a form which need not upset anybody's moral sensibilities. It was meant, in fact, to bring together the two ideals which lay behind the moral positions animating these two contentions – a realistic respect for science on the one hand and an enlightened enthusiasm for freedom on the other. So it attempted to do what I have been suggesting here is the chief business of philosophy – to bring together two genuine insights in a way that would make them both available to guide our lives.

The first result of this was, of course, confusion. A lot of cats got among a lot of pigeons, all the more awkwardly because prophets such as Robert Ardrey and Desmond Morris had lately written books about animals which blackened their reputations misleadingly and made it easier to reject comparison with them. But still, some of what I was saying, and what a number of other people were also saying at the time, did get across. Very gradually, the full-scale behaviourist and existentialist message did die down. As we shall see, it still haunts us, but it is no longer openly enforced by stern discipline.

Since that time, attitudes on these topics have certainly changed. A host of excellent ethologists, from Jane Goodall to David Attenborough, have educated us about animals in such a way that arguments concerning our continuity with them ought not to be needed any longer. Yet somehow it seems that they still are needed, and, more oddly still, that many people still see much of the rest of the non-human world as dispensable. In fact, our general attitude to nature has changed, but it has not moved very far. There is still a general impression that our species is something so special – so thoroughly cut off from the rest of nature, from what may be called 'the environment' – that, apart from the pleasures of the imagination, we need not really bother about all these other organisms.

Thus the 'environment' easily seems to become something of a luxury, perhaps like art – a spare-time occupation, something that should only be attended to after serious matters like the next budget have been properly settled. If it suited us (people sometimes feel) we could simply consume the lot on our own. This view is not, of course, often openly stated now, but for practical purposes our civilization has long been drifting towards it. Only very lately – perhaps only during the past few years? – do people with power to change the world, people with some money and influence, seem to be starting to notice that this attitude is not just unrealistic but, in the long run, certainly suicidal.

Nature-free thought

If we look back at the controversies that have occupied the scholarly world during this time, such as the clashes which surprised me so in the 1950s, we can see, first, that the clashing armies usually avoided attending to *nature* at all, and next, that they refused explicitly to attend to *human nature*.

On the first point, an example worth mentioning is water shortage. In Las Vegas and many other places, humans are now suffering from an alarming lack of water. Yet I recently heard of a scheme, tried out lately with success in Oregon and other places, whereby beavers, when they were allowed to build their own dams freely, raised the local water table (to the general surprise) far more effectively than human dam-builders could do it. And of course they did this without expecting to be paid for it.

This method will, of course, need to go through the usual exhausting series of tests before it can be widely admitted that nature, on this occasion, seems to have done at least as well as machinery. That delay will be needed because – as I keep explaining – our actions are not primarily ruled by evidence but much more by our imagery, our world-pictures, the myths that we have grown up accepting. In fact, as Wittgenstein said, 'a picture held us captive'.[1] And in our times, central to every myth – every mind-picture – is reliance on that all-purpose gadget, the Machine.

That is why even people who do worry seriously about human prospects often do not concentrate on thinking about how we might be able to behave more sensibly. They may prefer to rely on the development of machines, and they may hope that this development will be concentrated in the singularity. Of course this is understandable. Trying to alter human conduct is known to be very hard work, work which requires concentration and a real understanding of human nature. And there is still enough behaviourism in our mental cupboards to make that sort of understanding appear as an impossible ambition.

Yet what actually happens to us will surely still be determined by human choices. Not even the most admirable machines can make better choices than the people who are supposed to be programming them. So we had surely better rely here on using our own Minds rather than wait for Matter to do the job.

And, if this is right, I suspect that something else that was mentioned in those opening pages – namely, philosophical reasoning – will now become rather important. We shall need to think about *how* best to think about these new and difficult topics – how to imagine them, how to visualize them, how to fit them into a convincing world-picture. And if we don't do that for ourselves, it's hard to see who will be able to do it for us.

Notes

Chapter 2

1 Dummett, *Truth and Other Enigmas*, 458.

Chapter 3

1 Keats, 'On First Looking into Chapman's Homer', 1816.
2 Douglas, 'Can Our Knowledge of Biology Help Us Avoid Another Financial Meltdown?' 41.

Chapter 4

1 Carroll, 'It's Mind-Blowing what Our Puny Brains Can Do', 29.
2 Brooks, 'Quantum Entanglement Mangles Space and Time', 31.
3 Ibid.
4 Noble, *Dance to the Tune of Life*, 73.

Chapter 5

1   Huxley, *The Elements of Physiology and Hygiene*, 178.

Chapter 6

1   Huxley, *The Elements of Physiology and Hygiene*, 178.
2   Weisberg, 'The Hard Problem of Consciousness'.
3   McGinn, *The Making of a Philosopher*, 204.
4   McGinn, *The Mysterious Flame*, 14.

Chapter 7

1   Anderson, 'What Happened Before the Big Bang?'
2   Brooks, 'Don't Discard Any Quantum Options', 14.

Chapter 8

1   Chalmers, 'Why isn't there more progress in philosophy?' 4.
2   Ibid.
3   Ibid., 33.
4   Ibid., 5.
5   Tallis, 'The "P" word', 53.
6   Murdoch, *The Sovereignty of Good*, 34.

Chapter 9

1 Rousseau, *The Basic Political Writings*, 141.

Chapter 10

1 Mill, *Utilitarianism, Liberty and Representative Government*, 99.
2 Ibid., 106.

Chapter 13

1 1 Tim. 6.12.
2 Reginald Heber, The Son of God Goes Forth to War (1812).

Chapter 14

1 Krauss, unpublished talk.
2 *New Scientist*, 7 March 2015, front cover.
3 Wilson, *Sociobiology*, 575.
4 Wolpert, *The Unnatural Nature of Science*, 121.

Chapter 15

1 Pope, *The Works of Alexander Pope*, 114.
2 Crick, *The Astonishing Hypothesis*, 3, 7.

3 Searle, 'Minds, Brains and Programs', 368.

4 Conway Morris, *Life's Solution*, 127.

5 Ibid., 283–4.

6 Monod, *Chance and Necessity*, 165.

Chapter 16

1 Mill, 'On Liberty', 117.

2 Thornton, 'The Age of Machinery', 281.

3 Butler, *Erewhon*, 119.

4 Turing, 'Intelligent machinery, a heretical theory', 259–60.

Chapter 18

1 Turing, 'Computing machinery and intelligence', 442.

2 Dawkins, *The Selfish Gene*.

3 Xenopoulos, 'When paths diverge'.

4 Ibid., quoting Jason Silva (2014 seminar).

5 Kurzweil, *The Singularity Is Near*.

6 Good, 'Speculations concerning the first ultra-intelligent machine'.

7 Xenopoulos, 'When paths diverge'.

Chapter 19

1 McGinn, *The Making of a Philosopher*, 205.

2 Vinge, 'First Word', 10.

3 Moravec, *The Age of Robots*.

4 Butler, *Erewhon*, 128–9.

5 Kaczynski, *Industrial Society and Its Future*, 102.

Chapter 20

1 Austin Lee and Kimbriel, *The Resounding Soul*, 3.
2 Ibid.
3 Ibid., 354.
4 Ibid., 3.
5 Pinker, 'Thinking does not imply subjugating'.
6 *Leviathan*, Chapter 1, Section 5.
7 Ibid.
8 Crick, *The Astonishing Hypothesis*, 7.

Chapter 21

1 Pinker, 'Thinking does not imply subjugating'.
2 Robinson, *Absence of Mind*, 21–2.
3 Anon, *The Masque of Balliol*, 1881.
4 Anderson, 'What Happened Before the Big Bang?'

Chapter 22

1 Robinson, *Absence of Mind*, 7–8.
2 Midgley, *The Myths We Live By*, 91.

Chapter 23

1 Brockman, *What to Think About Machines That Think*.
2 Ibid., 85–7.
3 Ibid., 9–11.

Chapter 24

1 Nagel, *Mind and Cosmos*, 8–16.
2 Ibid., 66.
3 Kipling, 'In the Neolithic Age'.
4 Solomon, 'Literature About Medicine May Be All That Can Save Us', 3.
5 Robinson, *Absence of Mind*.

Conclusion

1 Wittgenstein, *Philosophical Investigations*, para. 115.

REFERENCES

Anderson, R. (2012), 'What Happened Before the Big Bang? The New Philosophy of Cosmology', *The Atlantic*, 19 January. Available online: https://www.theatlantic.com/technology/archive/2012/01/what-happened-before-the-big-bang-the-new-philosophy-of-cosmology/251608 (accessed 14 January 2018).

Austin Lee, E. and Kimbriel, S. (2015), *The Resounding Soul: Reflections on the Metaphysics and Vivacity of the Human Person*, Eugene, OR: Cascade Books.

Brockman, J. (2015), *What to Think About Machines That Think*, New York: Harper Perennial.

Brooks, M. (2015), 'Don't Discard Any Quantum Options', *New Scientist*, 14 November: 14.

Brooks, M. (2016), 'That's Odd: Quantum Entanglement Mangles Space and Time', *New Scientist*, 27 April: 31–2.

Butler, S. ([1872] 2012), *Erewhon*, North Chelmsford, MA: Courier Corporation.

Carroll, S. (2016), 'Opinion Interview: It's Mind-Blowing what Our Puny Brains Can Do', *New Scientist*, 16 April: 28–9.

Chalmers, D. (2015), 'Why isn't there more progress in philosophy?' *Philosophy*, 90 (1): 3–31.

Conway Morris, S. (2003), *Life's Solution: Inevitable Humans in a Lonely Universe*, Cambridge: Cambridge University Press.

Crick, F. (1994), *The Astonishing Hypothesis*, New York: Touchstone.

Dawkins, R. (2016), *The Selfish Gene: 40th Anniversary Edition*, Oxford: Oxford University Press.

Douglas, K. (2015), 'Can Our Knowledge of Biology Help Us Avoid Another Financial Meltdown? Slime-Mould Economics', *New Scientist*, 25 July: 38–41.

Dummett, M. (1978), *Truth and Other Enigmas*, London: Duckworth.

Good, I.J. (1965), 'Speculations concerning the first ultraintelligent machine', *Advances in Computers*, 6, 31–88.

Huxley, D.H. (1869), *The Elements of Physiology and Hygiene: A Text-Book for Educational Institutions*. New York: D. Appleton and Company.

Kaczynski, T.J. (2018), *Industrial Society and Its Future*, Pub House Books.

Krauss, L. (2014), Unpublished talk at 'Where the Light Gets In' festival, Hay-on-Wye, May.

Kurzweil, R. (2005), *The Singularity Is Near*, New York: Viking.

McGinn, C. (1999), *The Mysterious Flame*, New York, Basic Books.

McGinn, C. (2002), *The Making of a Philosopher*, New York: HarperCollins Publishers Inc.

Midgley, M. (2002), *Beast and Man: The Roots of Human Nature*, London: Routledge.

Midgley, M. (2004), *The Myths We Live By*, London: Routledge.

Mill, J.S. (2010), *Utilitarianism, Liberty and Representative Government*, edited by G. Williams, London: Everyman's Library.

Mill, J.S., ([1859] 1936), 'On Liberty', in *Utilitariamism, Liberty and Representative Government*, London: Dent and Dutton.

Monod, J. (1971), *Chance and Necessity: An Essay on the National Philosophy of Modern Biology*, Glasgow: A.A. Knopf.

Moravec, H. (1993), *The Age of Robots*. Available online: http://www.frc.ri.cmu.edu/~hpm/project.archive/general.articles/1993/Robot93.html (accessed 18 January 2018).

Murdoch, I. (1971), *The Sovereignty of Good*, New York: Schocken Books.

Nagel, T. (2012), *Mind and Cosmos: Why the Materialist Neo-Darwinian Conception of Nature Is Almost Certainly False*, Oxford: Oxford University Press.

Noble, D. (2017), *Dance to the Tune of Life: Biological Relativity*, Cambridge: Cambridge University Press.

Pinker, S. (2015), 'Thinking does not imply subjugating', *Edge*. Available online: https://philosophynow.org/issues/109/When_Paths_Diverge (accessed 18 January 2018).

Pope, A (1847), *The Works of Alexander Pope, Esq.*, London: Longman, Brown and Co.

Robinson, M. (2010). *Absence of Mind: The Dispelling of Inwardness from the Modern Myth of the Self*, New Haven, CT: Yale University Press.

Rousseau, J.-J. (1987), *The Basic Political Writings*, Indianapolis, IN: Hackett Publishing Company.

Searle, J. (1981), 'Minds, Brains and Programs', in D.R. Hofstadter and D. Dennett (eds), *The Mind's I*, 353–72, Brighton: Harvester Press.

Solomon, A. (2016), 'Literature About Medicine May Be All That Can Save Us', *Guardian Review*, 22 April: 3.

Tallis, R. (2016), 'The "P" word', *Philosophy Now*, 113: 52–3.

Thornton, R. (1847), 'The Age of Machinery', *The Primitive Expounder*, 4 (12 August): 281.

Turing, A. (1950), 'Computing machinery and intelligence', *Mind*, 49: 433–60.

Turing, A. (1996), 'Intelligent machinery, a heretical theory', *Philosophica Mathematica*, 3 (4): 256–60.

Vinge, V. (1983), 'First Word', *Omni*, January: 10.

Weisberg, J. (n.d.) 'The Hard Problem of Consciousness', Internet Encyclopedia of Philosophy. Available online: http://www.iep.utm.edu/hard-con/(accessed 24 January 2018).

Wilson, E.O. (1975), *Sociobiology*, Cambridge, MA: Harvard University Press.

Wolpert, L. (1992), *The Unnatural Nature of Science*, London: Faber and Faber.

Xenopoulos, J. (2015), 'When paths diverge', *Philosophy Now*, 109. Available online: https://philosophynow.org/issues/109/When_Paths_Diverge (accessed 17 January 2018).

INDEX